Exploring Time-Variation
in Survival Models

Inaugural–Dissertation
zur Erlangung des Grades Doctor oeconomiae publicae (Dr.oec.publ.)
an der Ludwig–Maximilians–Universität München

vorgelegt von

Ursula Berger

2001

Referent:	Prof. Dr. L. Fahrmeir
Koreferent:	Prof. Dr. K. Ulm
Promotionsabschlußberatung:	6. Februar 2002

Bibliografische Information Der Deutschen Bibliothek

Die Deutsche Bibliothek verzeichnet diese Publikation in der Deutschen
Nationalbibliografie; detaillierte bibliografische Daten sind im Internet über
http://dnb.ddb.de abrufbar.

ISBN 3-8325-0457-5

Logos Verlag Berlin
Comeniushof, Gubener Str. 47,
10243 Berlin
Tel.: +49 030 42 85 10 90
Fax: +49 030 42 85 10 92
INTERNET: http://www.logos-verlag.de

Vorwort

Die vorliegende Arbeit entstand während meiner Tätigkeit am Institut für Medizinische Statistik und Epidemiologie der Technischen Universität München. Die zugrundeliegende Projektarbeit wurde zu einem großen Teil im Rahmen des Sonderforschungsbereichs 386 von der Deutschen Forschungsgemeinschaft finanziert. Der Sonderforschungsbereich bot neben dieser Förderung wertvolle Möglichkeiten, die aktuellen Arbeiten auf Workshops und Konferenzen vorzustellen und mit anderen Wissenschaftlern zu diskutieren.

Mein besonderer Dank gilt den beiden Betreuern meiner Arbeit, Herrn Prof. Dr. Ludwig Fahrmeir und Herrn Prof. Dr. Kurt Ulm für ihre stete Unterstützung. Moreover, I'm grateful to Prof. Murray Aitkin. Whenever we met, he had time for inspiring discussions on Bayes factors and posterior Bayes factors. Herrn Prof. Dr. Stefan Klasen danke ich für die gute und unkomplizierte Kooperation bei der Studie zur Kindersterblichkeit in Afrika. Frau Dr. Nadia Harbeck und Herr Dr. Hjalmar Nekarda stellten mir freundlicherweise die Mammakarzinom- und die Magenkarzonom-Daten zur Verfügung.

Auch möchte ich mich bei allen Kolleginnen und Kollegen bedanken, die zum Gelingen dieser Arbeit beigetragen haben, insbesondere bei Juliane Schäfer für ihre zuverlässige Unterstützung und Martina Müller für ihre motivierenden Worte. Herrn Dr. Felix Dannegger danke ich für seine geduldige Hilfe mit der englischen Orthographie und Grammatik.

Rückblickend erwies sich die Forschung an zwei so kontroversen Ansätzen wie der Likelihoodbasierten Theorie und der Bayesianischen Statistik als eine besonders spannende und herausfordernde Aufgabe. Sie gab oftmals Anstoß für rege Diskussionen. An dieser Stelle möchte ich nicht zuletzt auch Göran für die zahlreichen anregenden Gespräche und hilfreichen Kommentare danken.

Contents

Chapter 1

Introduction

"Things should be made as simple as possible, but not any simpler."
Albert Einstein

The analysis of survival data has a longstanding tradition with applications in a wide range of different fields. For example, in the medical context it might be of interest to compare the survival of patients receiving different treatments, or in oncology one might want to explore the development of a disease dependent on individual characteristics; in epidemiology mortality of people might be studied with respect to their living conditions, while in economics the survival of newly founded companies may be regarded or the duration until a bank credit is fully paid back. The common focus of studies in these areas is to model duration time until the occurrence of a specific event. A particular interest is thereby to explore the impact of different factors or covariates. Classical survival models are thereby usually based on parametric assumptions, which typically imply that effects of covariates are constant over time. However, since survival analysis is based on observations on individuals over time, the restriction to time-constant effects appears somewhat artificial and one can easily imagine situations where the effects of covariates change over time. Reconsidering the examples from above, in oncology the impact of a tumour-related factor on survival might be small in the beginning of the disease but increase afterwards. Or when studying child mortality one might suspect that the importance of maternal characteristics may vary at different age phases of a child. Recent research on survival analysis has drawn its attention on the extension of the linear and parametric models to more flexible modelling methods in order to improve description of complex real worlds coherences, where a particular focus is on methods, which allow to incorporate dynamic effect structures.

It is obvious that while dynamic models are more flexible and improve goodness of a fit, allowing for time-variation will increase the model's dimension. Therefore an important issue, which should be kept in mind, is to keep the model as simple as possible. This boils down to a trade-off between goodness of fit and complexity of the model, comprehended as model selection.

This work concentrates on the identification and estimation of dynamic effects in different modelling frameworks. Two contrary concepts of statistics are considered: Likelihood based statistics and Bayesian theory. While model estimation in the likelihood framework is realised by maximising variants of the likelihood, in Bayesian statistics inference is based on posterior distributions of the model parameters. In the likelihood framework model selection can be based on significance testing, whereas within the Bayesian framework model selection demands for different methods. Both approaches will be developed side by side in this work. Besides providing a smooth fit for dynamic effects, a main focus of this work will be on model selection. This especially means to distinguish between a dynamic and a constant fit. The ultimate goal is to select a model that is as complex as needed and as parsimonious as possible. Simplicity in modelling not only implies the selection of a parsimonious predictor, but also concerns modelling techniques. The work will therefore concentrate on methods that are straightforward to use in practice and ensure feasible computational effort.

Chapter 2 gives a brief introduction to survival analysis. A special focus is thereby on the different models under investigation in the subsequent chapters. In practical survival analysis the most popular model is the proportional hazards regression model introduced by Cox (1972). Due to its simple structure and flexible definition it offers a suitable tool for many applications. Due to its wide acceptance the Cox model is implemented in most statistical packages. Nevertheless the fundamental assumption of the Cox model is the proportionality of the hazards, which implies time-constant effects. To assure acceptance of an extension in application it seems essential to maintain the advantages of the Cox model, that is to base the extension on methods which are straightforward to implement and use in practice. Moreover it should be ensured that methods are computationally feasible and easy to understand. Chapter 3 will present an extension of the Cox model using a flexible method based on Fractional Polynomials. This enables one to detect and simultaneously model dynamic effect structures within the Cox framework. The definition of Fractional Polynomials ensures a flexible fit by a large variety of possible shapes, including linear, bounded and asymptotic shapes. A likelihood ratio test will be presented that allows for the decision whether the effect of a covariate on the outcome can be assumed to be constant over time, or if it rather follows a dynamic structure. Overall, this can be regarded as a test on the proportional hazard assumption. Due to their flexibility, Fractional Polynomials allow to identify even rather complex departures from the PH-assumption. At the same time it provides a good fitting alternative when significant time-variation exists. Moreover, the approach preserves the linear structure of the predictor, and hence, implementation and inference is straightforward using standard estimation techniques. An algorithm will be described, that allows for fitting multivariate, semi-dynamic Cox models. In a simulation study the performance of the proposed test will be examined and compared to other PH-tests. The Fractional Polynomial approach will be employed to explore the prognostic impact of different tumour-related factors on survival of gastric cancer patients.

Chapter 4 and Chapter 5 focus on Bayesian approaches to dynamic survival modelling and model selection. Bayesian modelling has become more and more attractive in the last decade. One reason for this is that it allows to incorporate prior knowledge. But also the development of Markov Chain Monte Carlo methods and the increasing computational power enables one now to calculate flexible, high dimensional models mirroring complex structures. To decide on the effect structures of the covariates and to select a model that optimally trades off between model fit and model complexity, model criteria are required, which allow for the comparison of competing different Bayesian models of different dimensions. The Cox model assumes that survival time is measured continuously. In numerous applications, however, survival time is observed on a discrete scale only. In this case one can employ the logit model, which is here considered within the Bayesian framework. The analysis of discrete survival time data within the Bayesian framework is described in Chapter 4. The dynamic logit model will be employed together with a discrete smoothness prior. Inference of the approach relies on Markov Chain Monte Carlo methods and an algorithm will be outlined that successively generates posterior samples for all model parameters. The main focus of this chapter will be on approaches to compare different models and decide between dynamic and time-constant effects.

Besides the classical Bayes factor, alternative proposals will be investigated and compared. In particular, practical feasibility of the different proposals will be of major interest. This is regarded with respect to

• the comparison of complex hierarchical models of arbitrary dimensionality fitted by Markov Chain Monte Carlo sampling and

• the application in survival analysis, where additional problems arise from the particular data structure.

The behaviour of the different criteria will be explored in a comparative simulation study. This is of special interest as until now only little work has been published on comparing different Bayesian proposals. A data example investigating the determinants of infant mortality will provide an insight into the performance of the different Bayesian criteria when applied in a complex model selection process. In Chapter 5 the background of the different concepts will be restudied and illuminated in more depth. Associations between the criteria will be revealed.

Finally, in Chapter 6 both approaches, that is the Fractional Polynomials and the Bayesian logit modelling, will be contrasted in an analysis of breast cancer data. This chapter intends to give an 'objective' comparison of the two concepts.

Chapter 2

Modelling Survival Data

Survival analysis deals with the modelling of data from a population of individuals for whom time to the occurrence of some event, e.g. death, is treated as the outcome variable of interest. If this event is not death but some other outcome, survival time analysis is also more generally denoted as *failure time* or *duration time* analysis. Survival data or failure time data can come from a wide range of different origins. For example, in the medical context it might be of interest to compare the survival of patients receiving different treatments, in epidemiology one might study mortality of people with respect to their living conditions, while in economics the survival of newly founded companies may be regarded or the duration until a bank credit is fully paid back. In a demographic study the event might be the change of residence or in a botanical study the duration until a plant is blooming for the first time. Accordingly to the area of study very specific notations for duration time are sometimes used which refer to the type of event, e.g. *disease free survival* when observing time to relapse of a disease. In survival analysis the event of interest is *absorbing*, i.e. it terminates observation. One may also think of situations where the event of interest is not absorbing but may recur, such as an illness, or where more than one type of possible events may occur, such as several causes of death. Recurrent event models and competing risk models deal with this type of data. They are derived as extensions of survival models but are not within the scope of this thesis. For their description see e.g. Fahrmeir & Tutz (2001) or Cox & Oakes (1984).

Since the seminal publication of Cox (1972) considerable progress has been made in developing flexible methods for fitting survival data. A thorough introduction to methods for survival analysis is given by Kalbfleisch & Prentice (1980). Also Cox &

Oakes (1984) give a concise account of the analysis of survival data. They mainly focus on continuous survival time, for which they present a competitive overview of different continuous distributions. Fleming & Harrington (1991) and Andersen et al. (1991) approach this field from the view of counting processes. Survival models for discrete time and their extensions are found in Fahrmeir & Tutz (2001). For an application oriented introduction see e.g. Parmar & Machin (1996) or Harrell (2001).

In this chapter a short overview to survival analysis is given, with the emphasis on introducing the models and notation used in the following chapters.

2.1 Survival Times and Survival Data

2.1.1 Distribution of Survival Time

A principal task in survival modelling is to assess the dependency of failure time distributions on explanatory variables. These failure times might be continuous or discrete.

Continuous Survival Time

Consider survival time of a homogeneous population of individuals to be represented by a continuous, non-negative random variable T. Its density function is defined by

$$f(t) = \lim_{\Delta t \to 0^+} \frac{P\big(t \leq T < t + \Delta t\big)}{\Delta t} \, .$$

The function

$$S(t) = P(T \geq t) \qquad 0 < t < \infty$$

is the *survivor function* and gives the probability that an individual survives at least until time t. The principal function used to describe the distribution of survival time is the *hazard function*, describing the immediate 'risk' of failure for an individual known to be alive at time t. For continuous survival time is defined as the limit

$$\lambda(t) = \lim_{\Delta t \to 0^+} \frac{P\big(t \leq T < t + \Delta t | T \geq t\big)}{\Delta t} \, .$$

The hazard function is usually considered when comparing survival behaviour of different populations. It is also a convenient concept for defining models for data containing censoring (see Section 2.1.2).

Any of these functions, the density function, the survival function or the hazard function, fully specify the distribution of T. Their relationships are given by

$$f(t) = \lambda(t)S(t)$$

and

$$S(t) = \exp\left(-\int_0^t \lambda(u)du\right).$$ (2.1)

Discrete Survival Time

Depending on the type of application, failure time data may also be measured on a discrete scale, either due to grouping of continuous failure times, due to imprecise measurements or because time itself is discrete. For example, failure time may be only recorded in months or it may denote the number of attempts needed before experiencing success to perform a certain task. If failure time results from grouped continuous survival time, it is also called *interval censored* survival time.

Consider now the survival time T to be a discrete random variable taking values $t_1 < t_2 < \ldots$ with the probability function for failure at t being

$$f(t) = P(T = t).$$

For grouped continuous data, t indicates the interval $[a_{t-1}, a_t)$ and $T = t$ denotes failure within this interval. The discrete survivor function is given by the probability of reaching time point or interval t

$$S(t) = P(T \geq t)$$

and the discrete hazard function is defined as the conditional probability of failure at time t given that t is reached, i.e.

$$\lambda(t) = P(T = t|T \geq t).$$

Note that alternatively one may also consider the survival function $S^*(t) = S(t+1)$, which is the probability of surviving time t (Fahrmeir & Tutz, 2001).

In the discrete case, the relations between these three probabilities result in

$$S(t) = \prod_{k=1}^{t-1} \left(1 - \lambda(k)\right)$$

and

$$f(t) = \lambda(t)S(t) = \lambda(t)\prod_{k=1}^{t-1} \left(1 - \lambda(k)\right). \tag{2.2}$$

2.1.2   Censoring

A characteristic feature of survival data is the occurrence of censored information, that is, incompletely observed survival time for some of the data. Censoring is one of the major reasons that makes the field of survival analysis 'non standard' and the accommodation of incomplete survival time in the analysis requires specialised models and methods. *Right censored* survival time arises when the duration of observation terminates before an event has occurred so that only a 'truncated' survival time is available, which is known to be exceeded by the 'true' survival time. [1] In long-term studies, right censoring often naturally occurs if the reporting period is fixed in advance and ends before all individuals experienced the event of interest. In addition, individuals may randomly enter or exit a study over time and therefore have varying observation periods. These situations are common in medical studies, such as oncological studies as regarded in Chapter 3 and Chapter 6, where patients are followed-up after surgery on a tumour. Censoring also emerges if data on previous event history is collected at a certain time-point from a general sample including some individuals who never had an event. This is the case for the study of infant mortality discussed in Chapter 4, where data were collected in one-time interviews. One may also face more complex censoring mechanisms. References to different types of censorings are found in Kalbfleisch & Pentrice (1980) and Andersen et al. (1993).

---

[1] Accordingly, *left censored* survival time is truncated at its beginning, when the true starting point has not been observed. It involves a special data situation and requires different handling, (see e.g. Kalbfleisch & Pentrice, 1980, Appendix).

To incorporate information about censoring, survival is recorded by (T_n, δ_n), for $n = 1, \ldots, N$, where T_n denotes the observed survival time of individual n and δ_n is the indicator of failure, which takes the value 1 if individual n failed, and zero if it is censored. Thus, if T_n^* denotes the true survival time of individual n, δ_n is given by

$$\delta_n = \left\{ \begin{array}{ll} 0, & T_n < T_n^* \\ 1, & T_n = T_n^*. \end{array} \right.$$

A crucial assumption for censoring is that it is non-informative. This means that conditional on its population, the prognosis for any individual who has survived to T_n and is then censored, should not be affected by censoring, nor should censoring affect the survival of the remaining population. Violation of this assumption, for example in the presence of informative dropouts, requires additional modelling efforts.

2.1.3 The Likelihood Function

To link the distribution of survival time T to a set of explanatory variables x, regression models are introduced, which define the hazard function, the survival function and the probability/density function conditionally on the covariate values, i.e. $\lambda(t|x)$, $S(t|x)$ and $f(t|x)$. In this section general definitions of the likelihood functions of such regression models are derived for continuous as well as discrete survival time.

Continuous Survival Time

Consider first the case of continuous survival time. An individual n with covariates x_n, which fails at time t_n, contributes the term $f(t_n|x_n)$, to the likelihood, that is the density of failure at t_n. Assuming that censoring is non-informative and in particular not influenced by the parameters of interest, an individual m whose survival time is censored at t_m has the contribution $S(t_m|x_m)$, that is the probability of survival beyond t_m. The likelihood for N independent individuals is then given by

$$l = \prod_{n=1}^{N} f(t_n|x_n)^{\delta_n} S(t_n|x_n)^{1-\delta_n},$$

where δ_n is again denoting the survival index as defined above. Applying the relations (2.1), the likelihood can be re-written as

$$
\begin{aligned}
l &= \prod_{n=1}^{N} \lambda(t_n|x_n)^{\delta_n} S(t_n|x_n) \\
&= \prod_{n=1}^{N} \lambda(t_n|x_n)^{\delta_n} \exp\left(- \int_0^{t_n} \lambda(u)du \right).
\end{aligned}
\tag{2.3}
$$

Discrete Survival Time

In analogy to continuous survival time, also for discrete survival time the likelihood contribution of an individual n is

$$
l_n \propto f(t_n|x_n)^{\delta_n} S(t_n|x_n)^{1-\delta_n} = P(T_n = t_n|x_n)^{\delta_n} P(T_n \geq t_n|x_n)^{1-\delta_n}.
$$

Inserting the relation of the discrete probability functions (2.2) results in

$$
l_n \propto \lambda(t_n|x_n)^{\delta_n} \prod_{k=1}^{t_n-1} \left(1 - \lambda(k|x_n)\right).
$$

It is supposed that an individual censored at time t contributes only the information that its underlying survival time is known to exceed t. Nothing further is known. If continuous data are grouped and t denotes the interval $[a_{t-1}, a_t)$, this is equivalent to assuming that censoring occurs at the very end of the interval. Hence, if R_t denotes the set of individuals at risk at time t, this risk set also includes all individuals, which are censored in t, i.e. $R_t = \{n : T_n \geq t\}$. Alternatively, one may also suppose censoring at the beginning of an interval, as in Fahrmeir & Tutz (2001), or within the interval, as in Thompson (1977), by adjusting the risk set.

Discrete time survival models can be cast into the framework of binary regression models by considering for each individual its *survival experience* through time. This can be written as a process $y_n(t)$ which is defined as long as individual n is under risk and takes values

$$
y_n(t) = \left\{ \begin{array}{ll} 0, & t < T_n \\ \delta_n, & t = T_n \end{array} \right.
$$

for $t = 1, \ldots, T_n$. Thus, the n-th individual experiences a sequence of censorings at $t_1, t_2 \ldots$, and either failure or a final censoring in T_n, i.e. $y_n = (y_n(1), \ldots, y_n(T_n)) =$

$(0, \ldots, 0, \delta_n)$. In other words, for each time point t and all individuals still at risk, $y_n(t)$ can be regarded as a binary response following a Bernoulli distribution with $\lambda(t|x_n)$ giving the 'probability of success'. The likelihood contribution of a single individual is then the product over all observations points

$$l_n \propto \prod_{t=1}^{T_n} \lambda(t|x_n)^{y_n(t)} \big(1 - \lambda(t|x_n)\big)^{1-y_n(t)}$$

and the product over all individuals $n = 1, \ldots, N$ yields the full likelihood. It can be rewritten as

$$l \propto \prod_t \prod_{n \in R_t} \lambda(t|x_n)^{y_n(t)} \big(1 - \lambda(t|x_n)\big)^{1-y_n(t)},$$

multiplying over all time points t and risk sets R_t.

2.2 Regression Models for Continuous Survival Time

Various regression models have been proposed to fit continuous survival data. The differences between them are mainly based on the different assumptions made for the distributions of survival time and the particular link between survival time and a number of explanatory variables x. Common examples of continuous survival time models are the Exponential and the Weibull models, which are based on the corresponding distributions. In this thesis the focus is on a very general model for continuous survival time which completely avoids the specification of a survival distribution, that is the proportional hazards model proposed by Cox (1972).

2.2.1 The Cox model

The most popular model for fitting failure time data, especially in biometrical practice, is the proportional hazards regression model proposed by Cox (1972). Due to its flexible definition it offers a suitable tool for many applications and is particularly advantageous in studies where the focus lies on exploring the impact of a covariate on the hazard, rather then on estimating the hazard function itself. For ease of notation, assume that inference about a single covariate is of interest. The Cox model defines the hazard function as

$$\lambda(t|x) = \lambda_0(t) \exp(\beta x).$$ (2.4)

$\lambda_0(t)$ denotes the *baseline hazard* representing a general pattern of the hazard for failure. It is assumed to be the same for all observations. In contrast to other models, in the Cox model the baseline hazard can be any unknown (non-negative) function of time and remains unspecified. Only the regression coefficient β is determined, expressing the effect of the covariate on failure. The advantage of the Cox model lies in this 'semi-parametric' character, which ensures high flexibility while retaining a simple form. The factor $\exp(\beta x)$ is interpreted as the 'Relative Risk' as it represents the impact of covariate x that multiplicatively effects the hazard of failure.

When the baseline hazard is specified by $\lambda_0(t) = \lambda_0$ or $\lambda_0(t) = \lambda_0 p(\lambda_0 t)^{p-1}$ the parametric Exponential model or the Weibull model, respectively, arise as special cases. Non-parametric estimates of the baseline hazard are suggested for example by Cox (1972) and Breslow (1974).

Partial Likelihood Estimation

Estimation of the coefficient β in (2.4) is usually based on the partial likelihood approach introduced by Cox (1972, 1975). Consider that for K individuals, $K = \sum_{n=1}^{N} \delta_n$, distinct failure times $t_{(1)} < t_{(2)} < \ldots < t_{(K)}$ have been observed and the remaining $N - K$ observations are right censored. Let $R(t_{(k)})$ denote the set of individuals at risk just prior to $t_{(k)}$ and let $x_{(k)}$ be the value of x for the individual failing at $t_{(k)}$. Moreover, let x_l be the covariate value of the lth individual still at risk at time $t_{(k)}$, i.e. $l \in R(t_{(k)})$. The conditional probability that individual k fails at $t_{(k)}$ given that the individuals $R(t_{(k)})$ are at risk is

$$\frac{\lambda(t_{(k)}|x_{(k)})}{\sum\limits_{l \in R(t_{(k)})} \lambda(t_{(k)}|x_l)} = \frac{\exp(\beta x_{(k)})}{\sum\limits_{l \in R(t_{(k)})} \exp(\beta x_l)}.$$

The product over all failure points $t_{(1)}, \ldots, t_{(K)}$ defines the *partial likelihood*

$$pl(\boldsymbol{\beta}) = \prod_{k=1}^{K} \frac{\exp(\beta x_{(k)})}{\sum\limits_{l \in R(t_{(k)})} \exp(\beta x_l)}.$$ (2.5)

To illustrate why (2.5) is a 'partial' likelihood, consider the full likelihood function of the Cox model. By using (2.3) and (2.4) it can be written as

$$
\begin{aligned}
l(\boldsymbol{\beta}) \;&\propto\; \prod_{n=1}^{N} f(t_n)^{\delta_n} S(t_n)^{1-\delta_n} \\
&=\; \prod_{n=1}^{N} \lambda(t_n)^{\delta_n} S(t_n) \\
&=\; \prod_{n=1}^{N} \Big(\lambda_0(t_n) \exp(\beta x_n) \Big)^{\delta_n} \exp \Big(- \int_0^{t_{(n)}} \lambda_0(u) \exp(\beta x_n) du \Big).
\end{aligned}
$$

Multiplying and dividing the sum over the risk set $R(t_{(k)})$ results in

$$
l(\boldsymbol{\beta}) \;\propto\; \prod_{k=1}^{K} \frac{\exp(\beta x_{(k)})}{\sum\limits_{l \in R(t_{(k)})} \exp(\beta x_l)} \cdot
$$

$$
\sum_{l \in R(t_{(k)})} \lambda_0(t_{(n)}) \exp(\beta x_l) \cdot \prod_{n=1}^{N} \exp \Big(- \int_0^{t_{(n)}} \lambda_0(u) \exp(\beta x_n) du \Big)
$$

where the first term does not depend on λ_0 and defines the 'partial' likelihood. The partial likelihood is based on the fact that $\lambda_0(t)$ is completely unspecified and hence no additional information can be contributed by the time intervals $(t_{(k-1)}, t_{(k)})$ in which no failures occurred. In these intervals the hazard of failure is assumed to be constant.

Maximising (2.5) gives maximum partial likelihood estimates that are shown to be consistent and asymptotically normally distributed. As an alternative to partial likelihood estimation Kalbfleisch & Pentrice (1980) derive estimates from the marginal distribution of the ranks of the failure times.

In definition (2.5) it is supposed that the K failures occur at K distinct time points. Even if survival time is thought of as a continuous variable, the preciseness of measuring survival time is limited and *ties* may result, i.e. two or more individuals may have identical observed survival times. Then $K < \sum_{n=1}^{N} \delta_n$, and the partial likelihood must be adjusted for these tied observations. The most common adjustment of the partial likelihood in the presence of ties was proposed by Breslow (1974):

$$
pl(\boldsymbol{\beta}) = \prod_{k=1}^{K} \frac{\prod\limits_{l \in D_k} \exp(\beta x_{(l)})}{\Big(\sum\limits_{l \in R(t_{(k)})} \exp(\beta x_l) \Big)^{d_k}}, \tag{2.6}
$$

where D_k is the set of d_k individuals failing at $t_{(k)}$. A more precise approximation correcting the denominator for multiple counting of failed individuals is given by Efron (1977). (See also Cox & Oakes, (1984).)

Ties might also occur for censored survival times which then affects the risk set. For continuous survival models when censoring coincides with a failure at time t, it is usually agreed that censoring occurs *after* the event and the risk set R_t also includes all individuals that are censored in t.

2.2.2   The Proportional Hazards Assumption

In model definition (2.4) all hazard functions, which are obtained for a specific covariate value are a multiple of the baseline hazard. Models that have this property are called *proportional hazards models* and the underlying assumption is called the *proportional hazards assumption* (PH-assumption). The property of proportionality becomes apparent when regarding the ratio of the hazard functions of two subpopulations with covariate values x_0 and x_1:

$$\frac{\lambda(t|x_1)}{\lambda(t|x_0)} = \frac{\exp(\beta x_1)}{\exp(\beta x_0)} = \exp\left(\beta(x_1 - x_0)\right). \tag{2.7}$$

This ratio only depends on the difference $x_1 - x_0$, but not on the level of x_0, and in particular it does not depend on time. Or more simply, if $x_1 - x_0 = 1$, i.e. x is increased by one unit, (2.7) yields in $\exp(\beta)$, which gives the factor by which the hazard of failure changes with x. Within PH-models this factor is assumed to be constant over time. Thus, (2.7) implies that the impact of a covariate measured at a baseline time-point (e.g. birth, surgery, first diagnosis, beginning of a treatment) is expected to remain unchanged over the observation period. It is obvious that this condition is rather stringent and questionable in numerous situations.

Hence, even though the Cox model is very flexible and suitable for many applications, there are considerable limitations which demand for an extension of the model. One extension is to let the baseline hazard vary in specific subsets or 'strata' of the data, resulting in the *stratified Cox model*. Another important generalisation is to allow for time-dependent covariates. Finally, one may relax the proportional hazards assumption by allowing the covariate effects to vary over time, i.e. by including dynamic effects. This latter extension takes into account that some covariates might have changing impact on failure at different periods of survival. The extension of the Cox model to accommodate dynamic effect structures is treated in detail in Chapter 3, where a flexible method is presented to explore time-variation. Since this method will make use of the other two generalisations, they are briefly described in the following sections.

2.2.3 The Stratified Cox Model

A possible extension of the Cox PH-model is to allow the shape of baseline hazard $\lambda_0(t)$ to vary in specific subsets of the sample. Suppose that the sample is divided into S strata, e.g. gender, each containing K_s distinct failure points. The hazard of the sth stratum can then be written as

$$\lambda_s(t|x) = \lambda_{0s}(t) \exp(\beta x)$$

where the coefficient β applies to all strata. For each stratum the baseline hazard $\lambda_{0s}(t)$ is assumed to be an arbitrary function, which is completely unrelated to the baseline of the other strata. Let $x_{(k^*s)}$ be the value of covariate x for the individual from stratum s failing at $t_{(k^*)}$. Assume again that survival time is measured in continuous time, precluding ties. The partial likelihood of the stratified model combines all strata and has the form

$$pl(\beta) = \prod_{s=1}^{S} \prod_{k^*=1}^{K_s} \frac{\exp(\beta x_{(k^*s)})}{\sum\limits_{l \in R_s(t_{(k^*s)})} \exp(\beta x_l)} \, . \tag{2.8}$$

2.2.4 The Cox Model for Time-Dependent Covariates

A second generalisation allows for time-dependent covariates $z(t)$, which might e.g. result from repeated measuring of the same variable. Two classes of covariates are distinguished here, *external* and *internal* covariates. An external covariate is not directly involved with the failure mechanism, as e.g. for a defined covariate, whose full path $\mathbf{Z} = z(1), z(2), \ldots$ is set from the beginning of the study for each individual. It can directly be incorporated in the hazard function by

$$\lambda(t|\mathbf{Z}) = \lim_{\Delta t \to 0^+} P\big(t \leq T < t + \Delta t | T \geq t, \mathbf{Z}\big) / \Delta t \, .$$

An internal covariate denotes the output of a stochastic process, which is generated by the individual during the study. It is observed only as long as the individual survives or is uncensored and hence carries information about survival time. The hazard function is therefore defined such that it conditions on the covariate history $\mathbf{Z}(t)$ up to time t but not further, i.e.

$$\lambda(t|\mathbf{Z}(t)) = \lim_{\Delta t \to 0^+} P\big(t \leq T < t + \Delta t | T \geq t, \mathbf{Z}(t)\big)/\Delta t.$$

The partial likelihood of the model with a time-dependent covariate is given by

$$pl(\boldsymbol{\beta}) = \prod_{k=1}^{K} \frac{\lambda\big(t_{(k)}|\mathbf{Z}_{(k)}(t_{(k)})\big)}{\sum\limits_{l \in R(t_{(k)})} \lambda\big(t_{(k)}|\mathbf{Z}_l(t_{(k)})\big)}$$

where $\mathbf{Z}_{(k)}$ is the available covariate history for the individual failing at $t_{(k)}$ and $\mathbf{Z}_l$ is the covariate path for the lth individual of the risk set $R(t_{(k)})$. Regarding the usual case, for which the hazard at t does not depend on the full covariate history $\mathbf{Z}(t)$ but only on the current value $z(t)$, the partial likelihood reduces to

$$pl(\boldsymbol{\beta}) \;\;=\;\; \prod_{k=1}^{K} \frac{\exp\big(\beta z_{(k)}(t_{(k)})\big)}{\sum\limits_{l \in R(t_{(k)})} \exp\big(\beta z_l(t_{(k)})\big)} \tag{2.9}$$

When comparing this definition with the partial likelihood of the stratified Cox model, it is easily seen that (2.9) can be transformed to (2.8) if the distinct failure points $t_{(1)}, \ldots, t_{(K)}$ are used for stratification, so that each risk set $R(t_{(k)})$ forms a stratum, and the covariates in the stratum are set to $x_l = z_l(t_{(k)})$ and $x_{(k^* s)} = z_{(k)}(t_{(k)})$, respectively. Then each stratum includes exactly one failure so that $K_s = 1$ and $S = K$, and (2.8) now becomes (2.9). Note that this connection makes use of the fact that the partial likelihood approach assumes the temporal change of the covariate to follow a step function.

2.2.5   The Dynamic Cox Model

An important extension of the linear predictor βx, which provokes a relaxation of the PH-assumption, copes with dynamic effect structures. Since in survival analysis individual data are collected over time, the restriction to time-constant effects appears as an artificial constraint, and relaxing the model to allow for effects, which vary over time, is more compelling. This is achieved in the *dynamic* Cox model

$$\lambda(t|x) = \lambda_0(t) \exp\big(\beta(t)x\big). \tag{2.10}$$

Models of the form (2.10) are generally referred to as *Varying-Coefficient Models* by Hastie & Tibshirani (1993) but have appeared earlier as models with time-dependent effects, e.g. in the time series literature or in Cox (1972). The shape of $\beta(t)$ mirrors the interaction between the covariate and time. It is usually fitted by smooth functions, where an attractive method is offered by Fractional Polynomials studied in Chapter 3[2].

2.3 Regression Models for Discrete Survival Time

If continuous time is grouped into broad intervals $[a_{t-1}, a_t)$, $t = 1, 2, \ldots$ or if survival time is discrete with few time points, many ties will occur and the estimation of the Cox model using approximation (2.6) will be poor. In the following section, two models are described, which more suitably handle discrete survival time.

2.3.1 The Grouped Cox Model

The direct analogue of the proportional hazards model of Cox (2.4) for discrete survival time is the *grouped proportional hazard model*, which defines the hazard function as

$$\lambda(t|x) = 1 - \exp\big(-\exp(\gamma_t + \beta x)\big) \tag{2.11}$$

where the coefficient β describes the effect of covariate x on the risk of failure and where γ_t represents the baseline effect (Fahrmeir & Tutz, 2001). Thus the baseline hazard $\lambda_0(t)$, which is generally defined as the hazard at the covariate value zero, takes the form

$$\lambda_0(t) = \lambda(t|x = 0) = 1 - \exp\big(-\exp(\gamma_t)\big).$$

---

[2]Another important extension of the linear predictor is to abandon linearity by allowing effects $\beta(x)$ to depend on the actual level of x. This yields a model definition of the class of *Generalized Additive Models*. It is, however, not within the scope of this work. For an introduction see e.g. Hastie & Tibshirani, 1990.

Alternatively, the grouped Cox model can be written in the multiplicative formulation

$$-\log\left(1 - \lambda(t|x)\right) = \exp(\gamma_t + \beta x) = \exp(\gamma_t)\exp(\beta x) \qquad (2.12)$$

where

$$\exp(\gamma_t) = -\log\left(1 - \lambda_0(t)\right).$$

To illustrate the relation between the continuous Cox model and the grouped Cox model consider the continuous survival time being partitioned into non-overlapping intervals $[a_0, a_1), [a_1, a_2), \ldots$, where t denotes the interval $[a_{t-1}, a_t)$ and $\Delta a_t = a_t - a_{t-1}$ is the length of the interval. Then the hazard function is $P(T = t | T \geq t, x) = P(a \leq T < a + \Delta a | T \geq t, x)$, abbreviating the t-th interval with $[a_{t-1}, a_t) = [a, a + \Delta a)$. Further let $\lambda^c(a|x)$ and $S^c(a|x)$ denote the continuous hazard function and survival function, respectively.

By expressing (2.12) in terms of the intervals of the continuous covariate a, using $S^c(a|x) = \exp(-\int_0^a \lambda^c(u|x)du)$ and $\lambda(t|x) = 1 - \exp(-\int_a^{a+\Delta a} \lambda^c(u|x)du)$, it is easily seen that for interval length $\Delta a \to 0$ one obtains the continuous Cox model $\lambda^c(a|x) = \lambda_0^c(a)\exp(\beta x)$. Thus, 'discretising' the Cox model does not influence the coefficient β, so that β and thus the relative risk parameter $\exp(\beta x)$ of the grouped Cox model are directly comparable to the corresponding estimates of the continuous Cox model.

It should be noted that the assumption for grouped continuous data that censoring occurs only prior to the end of an interval is deduced from the assumption that in continuous time censored observations at time t always follow any failure at t.

In principal the partial likelihood approach could be used to estimate the effect β. However, one would face the same difficulties as described above if many observations are tied, which will usually be the case for discrete or interval censored data. Alternatively, all parameters β and γ_t can be estimated by maximising the likelihood

$$l = \prod_{k=1}^{K}\left(\prod_{l \in D_k}\left[1 - \exp\left(-\exp(\gamma_k + \beta x_l)\right)\right] \cdot \prod_{l \in R_k} -\exp(\gamma_k + \beta x_l)\right)$$

where R_k again denotes the risk set at time point k, D_k is the set of individuals failing at k and K is the latest observed interval or time point. A problem with unconstrained maximum likelihood estimation for the baseline effects $\gamma_t, t = 1, 2, \ldots$ is that too many parameters have to be estimated. This can lead to inefficient or even non-existing estimates.

2.3.2   The Logit Model

An alternative model is the logistic model for discrete failure times as introduced by Thompson (1977). It directly emerges when regarding the individual survival experience $y_n = (y_n(1), \ldots, y_n(T_n)) = (0, 0, \ldots, \delta_n)$ described in 2.1.3. Here for each time point t and for those individuals still at risk, $y_n(t)$ is a binary response variable following a Bernoulli distribution with event probability $\lambda(t|x)$. Generally, this can be linked to the corresponding covariates x_n by a predictor $\eta_{nt}(x_n)$ and a response function $h : I\!R \to [0,1]$. The hazard function is then expressed through a binary response model

$$\lambda(t|x) = P(y(t) = 1 | T \geq t, x) = h(\eta_t).$$

Note that with this definition the introduction of time-dependent covariates is straightforward. In particular x may be replaced by the covariate history $\mathbf{X}(t) = x_1, x_2, \ldots, x_t$.

Using the logit-link to relate the predictor to the hazard function leads to the *logit model* for discrete survival time, where the hazard function is modelled by

$$\lambda(t|x) = \frac{\exp(\eta_t)}{1 + \exp(\eta_t)}. \tag{2.13}$$

In the simplest form, the predictor is $\eta_t = \gamma_t + \beta x$. Since this model is of the class of generalised linear models (Fahrmeir & Tutz, 2001), maximum likelihood estimates for the parameters γ_t and β may be calculated in the same way as in the general linear model framework. The likelihood contribution of one individual is given by

$$l_n = \prod_{t=1}^{T_n} \lambda(t|x_n)^{y_n(t)} \cdot \left(1 - \lambda(t|x_n)\right)^{(1-y_n(t))}. \tag{2.14}$$

With the recoding from above, the full likelihood function can then be written as

$$l = \prod_t \prod_{l \in R(t)} \lambda(t|x_l)^{y_l(t)} \cdot \left(1 - \lambda(t|x_l)\right)^{(1-y_l(t))}.$$

When the grouping intervals become short and contain no ties, the discrete logit model also leads back to the continuous Cox model (Thompson, 1977; Fleming & Harrington, 1991). Consider as before interval censored data of the continuous failure time a and let t denote the interval $[a_{t-1}, a_t)$. Furhter, let $\lambda^c(a|x)$ and $S^c(a|x)$ denote the continuous hazard function and survival function, respectively. The logit model implies that

$$\frac{\lambda(t|x)}{1 - \lambda(t|x)} = \exp(\gamma_t + \beta x)$$

and with $\lambda_0(t) = \lambda(t|x = 0)$

$$\frac{\lambda(t|x)}{1 - \lambda(t|x)} = \frac{\lambda_0(t)}{1 - \lambda_0(t)} \exp(\beta x) \tag{2.15}$$

Again with $S^c(a|x) = \exp\left(-\int_0^a \lambda^c(u|x)du\right)$ and $\lambda(t|x) = 1 - \exp\left(-\int_a^{a+\Delta a} \lambda^c(u|x)du\right)$ and abbreviating the t-th interval with $[a_{t-1}, a_t) = [a, a + \Delta a)$, relation (2.15) can be expressed in terms of continuous time by

$$\frac{1 - \exp\left(-\int\limits_a^{a+\Delta a} \lambda^c(u|x)du\right)}{\exp\left(-\int\limits_a^{a+\Delta a} \lambda^c(u|x)du\right)} = \frac{1 - \exp\left(-\int\limits_a^{a+\Delta a} \lambda_0^c(u)du\right)}{\exp\left(-\int\limits_a^{a+\Delta a} \lambda_0^c(u)du\right)} \cdot \exp(\beta x),$$

$$\Leftrightarrow \frac{1 - \exp\left(-\int\limits_a^{a+\Delta a} \lambda^c(u|x)du\right)}{1 - \exp\left(-\int\limits_a^{a+\Delta a} \lambda_0^c(u)du\right)} = \frac{\exp\left(-\int\limits_a^{a+\Delta a} \lambda^c(u|x)du\right)}{\exp\left(-\int\limits_a^{a+\Delta a} \lambda_0^c(u)du\right)} \cdot \exp(\beta x).$$

$$\tag{2.16}$$

Let now the interval length approaching zero, i.e. $\Delta a \to 0$. One can then apply l'Hôspital's rule to the left side of (2.16) using the derivative of the hazard function, which has the form

$$\frac{\partial}{\partial \Delta a}\left[1 - \exp\left(-\int\limits_0^{a+\Delta a} \lambda^c(u|x)du + \int\limits_0^a \lambda^c(u|x)du\right)\right]$$

$$= -\exp\left(-\int\limits_0^{a+\Delta a} \lambda^c(u|x)du + \int\limits_0^a \lambda^c(u|x)du\right) \cdot -\lambda^c(a + \Delta a|x)$$

$$= \exp\left(-\int\limits_a^{a+\Delta a} \lambda^c(u|x)du\right) \cdot \lambda^c(a + \Delta a|x),$$

which directly yields the proportional hazard model of Cox (2.4).

In principal, instead of (2.11) or (2.13) any other link of the type $h : \mathbb{R} \to [0, 1]$ could be used to relate the predictor and the hazard. A common alternative is for example the probit link, which provides useful properties that are of advantage for some model determination and estimation methods.

2.3.3 Dynamic Effects

For both discrete models the linear predictor in its simplest form is

$$\eta_t = \gamma_t + \beta x \,. \tag{2.17}$$

Here, in addition to the covariate effect β, for each time point $1, 2, \ldots, K$ for which the risk set is not empty the parameter γ_t has to be estimated. Assuming that $\gamma_1, \gamma_2, \ldots, \gamma_K$ are independent, this can be realised by maximising the likelihood for $(\beta, \gamma_1, \gamma_2, \ldots, \gamma_K)$. However, if continuous time is divided into many intervals or the number of time points K is large, the ratio of parameters to sample size is high and the resulting estimate will be poor or even non-existent. To overcome this problem an reduce model dimension, instead of estimating the baseline effect at every time-point independently, γ_t could be described by a function of time. In general it is difficult, if not impossible, to specify a parametric functional form for the baseline effect in advance, and again flexible estimation methods are required to explore its dynamic shape. These could be smooth estimation, which relates adjacent parameters via some presumed smoothness restrictions (Efron, 1988). The resulting estimate is a smooth function of time for the baseline effect γ_t. Predictor (2.17) can then be regarded as a *semi-parametric* predictor. Extending such a model to dynamic covariate effects requires just a small step by introducing time-varying coefficients $\beta(t)$ in the same way, i.e.

$$\eta_t = \gamma_t + \beta(t)x \,.$$

Estimation of the baseline effect γ_t and of the dynamic covariate effect $\beta(t)$ can be jointly based on the same smoothing approach. Here, in principle the same methods can be employed as for the Cox model, in particular Fractional Polynomials. However, since time is assumed to be discrete it seems more natural to apply methods for discrete effect modifiers.

In Chapter 4 a Bayesian approach is considered to fit discrete survival data with dynamic effect structures, using second-order random walk priors to attain smoothness. These can be regarded as the discrete analogue to second derivative penalties used for natural smoothing splines. A major focus in that chapter is Bayesian model choice with particular emphasis placed on the selection of those covariates which substantially affect survival and on the exploration if these effects distinctly vary over time.

Chapter 3

Dynamic Cox Modelling Based on Fractional Polynomials

This chapter focuses on extensions of the survival model proposed by Cox (1972) towards non-proportional hazards, where the exploration of the dynamic effect structures is accomplished within the partial likelihood framework. In practice, the Cox model is still the most popular model used for modelling failure time data. One reason for this is that estimation procedures for Cox's proportional hazard model are implemented in most statistical software packages. The intention of this chapter is therefore to design a method that allows for a flexible description of time-varying effects while taking advantage of standard estimation techniques and available procedures. This is achieved by dynamic Cox modelling based on Fractional Polynomials. By preserving the linear structure of the model the approach allows for a transformation of the dynamic predictor which leads back to a conventional Cox model. Hence fitting is straightforward using standard estimation techniques. Moreover, the approach offers the possibility to verify the existence of time-variation by applying a likelihood ratio test. This in turn allows to decide between dynamic and time-constant modelling.

The content of this chapter is based on Berger et al. (2003). In Section 3.1 a definition of Fractional Polynomials (FP) is given and their advantages in modelling dynamic effect structures are pointed out. An estimation procedure is described, which is based on the partial likelihood approach. Section 3.2 deals with model validation and model selection in order to identify temporal variation of the effect structure as well as to perform covariate selection. The approach is then further de-

veloped to an algorithm for fitting multivariate semi-dynamic models, which provides
a data-driven decision on the optimal dynamic structure and additionally allows for
iterative variable selection. The properties of the Fractional Polynomial approach
are explored in a comparative simulation study in Section 3.3, regarding its capability to identify dynamic structures. In Section 3.4 the FP-method is employed in an
analysis of gastric cancer data, where the objective is to give a careful description of
the prognostic effects of two new tumour-biological factors while adjusting for other
well established predictors. The chapter closes with a discussion of the FP method
summarising its advantages as well as its limitations.

3.1   The Dynamic Cox Model

For ease of notation and unless otherwise stated, it is assumed through out this
section that inference about a single covariate is of interest. To describe a dynamic
development of the relative risk, the Cox model can be extended for non-proportional
hazards as described in Section 2.2.2 by allowing the effect to vary with time

$$\mathbf{RR}(X, t) = \frac{\lambda(t|X)}{\lambda_0(t)} = \exp\left\{\beta(t)X\right\} \tag{3.1}$$

where the varying coefficient $\beta(t)$ expresses a particular kind of interaction with
time. The estimate of $\beta(t)$ is usually postulated to be smooth or at least not jagged.

A variety of methods have been suggested to estimate time-varying coefficients
within the Cox framework. A direct solution is to fit a piecewise constant model
in pre-specified disjoint time intervals, which results in a step function estimate of
$\beta(t)$. To obtain a smooth estimate, Hess (1994) suggests to substitute the constant
pieces of $\beta(t)$ by basis functions of cubic regression splines for a pre-specified grid
of knots. Hess shows, that under some restrictions such as linearity at both ends,
the cubic regression spline with κ knots can be written in the form

$$\beta(t) = \beta_0 + \beta_1 t + \sum_{j=1}^{\kappa-2} \beta_{j+1} z_j(t)$$

with $\kappa - 2$ terms $z_j(t)$ depending on time t and the locations of knots. Other authors
introduced smooth estimators using penalty functions combined with a smoothing
parameter which takes control of the trade-off between fit and roughness (e.g. Gray,
1992; Verweij & van Houwelingen, 1995). The most common approach in this class is

using natural smoothing splines as suggested by Hastie & Tibshirani (1993). Within the Cox framework smoothing splines are derived from maximising the penalised partial likelihood

$$l^p = l - \frac{1}{2}\tau \int \beta''(t)^2 dt \qquad (3.2)$$

where l is the partial likelihood measuring the fit, which is penalised by the squared second derivative of the effect function. The 'smoothing parameter' τ controls the trade-off between the fit l and the roughness measured by the second derivative. This results in a cubic spline function with knots at each failure time-point (Zucker & Karr, 1990). An alternative method is given by Grambsch & Therneau (1994), who propose to study scatter-plot smoothers of the scaled Schoenfeld residuals (Schoenfeld, 1982) against time. Also Bayesian approaches have been described, e.g. by Sargent (1997), who defines a hierarchical Cox model with state-space structure and uses MCMC-methods to estimate time-varying coefficients.

A main problem with the above proposals is that they demand pre-specifications, which noticeably influence the shape of the estimate $\hat{\beta}(t)$. This means, for example that the number and location of knots, a smoothing parameter or prior assumptions on the dynamic structure are required. More crucially however, most of these non-parametric methods require non-standard estimation techniques and their application is therefore not straightforward. Another equally important issue is the verification of the chosen structural form. The selection of an appropriate test-statistic for comparison of complex dynamic models with the PH-model is often not straightforward so that it is difficult to decide whether the improvement of the fit including time-variation justifies the increase in the model's complexity. However, to ensure that a model reaches acceptance in practice, it is important to keep it as parsimonious as possible and include time-variation only when evidence is given in the data.

An approach that makes use of ready available estimation techniques and allows for a direct verification of the dynamic fit was already described in Cox's original paper (Cox, 1972). Cox proposes to extend his model of proportional hazards towards

$$\lambda(t|x) = \lambda_0(t) \exp\left\{(\beta_0 + \beta_1 \varphi(t))x\right\}, \qquad (3.3)$$

that is he suggests adding some arbitrary transformation of time $\varphi(t)$ together with a further regression coefficient β_1 to the constant coefficient. This extension can be regarded as the introduction of a 'constructed' time-dependent covariate $Z(t) := \varphi(t)x$, constructed as an interaction term of the covariate X with time. For $\varphi(t)$ fixed, model (3.3) can therefore be determined via the partial likelihood estimation for time-dependent covariates (see Section 2.2.4). Moreover, testing for time-variation reduces to verifying $H_0 : \beta_1 = 0$, where common test statistics such as likelihood-ratio statistics can be used (Cox, 1972). Hence, this extension offers a simple way to investigate consistency with the PH-assumption and directly provides an alternative model when the assumption is rejected. However, the goodness of the fit and the power of the test for non-proportional hazards distinctly depends on the choice of the transformation $\varphi(t)$, where typically only simple constructions are used. Stablein et al. (1981) propose quadratic polynomials in order to model treatment effects that rise in the beginning and decrease later. Gore et al. (1984) employ exponential functions to describe the exponential decay of the relevance of certain covariates in a breast cancer study. Typically, the choices are less subtle, e.g. $\log(t)$, t or $rank(t)$, and hence less flexible. It is therefore a general objective to substitute $\varphi(t)$ by a more flexible function which provides an appropriate and smooth fit for $\beta(t)$ and can be generated by a data-driven algorithm. In addition the advantages of using standard estimation and testing techniques should be retained. This is achieved by Fractional Polynomial functions as described in the following section.

3.1.1   Fractional Polynomials

A Fractional Polynomial of degree m for any single continuous variable $t > 0$ is defined as

$$\phi_m(t, \mathbf{p}) = \beta_0 + \sum_{j=1}^{m} \beta_j t^{(p_j)}, \qquad (3.4)$$

where m is a positive integer, β_j are regression coefficients and $p_1 \leq \ldots \leq p_m$ are any positive or negative real-valued exponents. The logarithmic function is included for $p_j = 0$ by defining $t^{(0)} := \log t$. This definition of FPs was introduced by Royston & Altman (1994) and provides a wide range of flexible shapes. For $p_j \in I\!N$ the function ϕ_m is a conventional polynomial, while powers $p_j < 0$ render asymptotic shapes. The possibility of repeated powers $p_i = \ldots = p_k$, $i < k \leq m$ additionally involve combinations with $\log t$. For instance a FP of degree $m = 2$ with powers $\mathbf{p} = (2, 2)$ is defined as: $\phi_2(t, (2, 2)) = \beta_0 + \beta_1 t^2 + \beta_2 t^2 \log t$.

Originally Royston & Altman (1994) proposed Fractional Polynomials in order to model additive effect structures. Later Sauerbrei & Royston (1999) employed their additive FP approach to analyse the prognostic relevance of continuous factors for survival data. In this thesis Fractional Polynomial functions shall be proposed to model time-varying coefficients in the dynamic Cox model (3.1). By defining the time-varying effects to be $\beta(t) := \phi_m(t, \mathbf{p})$ a dynamic hazard function of the form

$$\lambda(t|x) = \lambda_0(t) \exp\left\{\phi_m(t, \mathbf{p})x\right\} = \lambda_0(t) \exp\left\{\left(\beta_0 + \sum_{j=1}^{m}\beta_j t^{(p_j)}\right)x\right\} \qquad (3.5)$$

is obtained. Thus Cox's proposal (3.3) is extended by adding m transformations of time $\varphi_1(t) = t^{(p_1)}, \ldots, \varphi_m(t) = t^{(p_m)}$. Since survival time t is positive, $t^{(p_j)}$ is well defined for any $p_j \in \mathbb{R}$. The flexibility of the FP approach is achieved by determining the degree of the Fractional Polynomial m and its powers $\mathbf{p}$ in accordance with the data at hand.

As seen in simulation studies and applications the definition of Fractional Polynomials is particularly suitable for fitting time-dependent effect structures in survival models. It contains the one-term transformations $\log(t)$, t or t^2 which have been used by many authors, as well as conventional polynomials. Moreover, in the dynamic modelling context, its asymptotic shapes obtained for $p_j < 0$ are of particular benefit when effects of covariates are reckoned to vanish with time. Complex time-dependencies arise when more than one term is included, i.e. $m > 1$. Thus, the degree of the Fractional Polynomial m directly controls the amount of complexity of the fit.

3.1.2 Model Estimation

It is immediately seen that the FP approach preserves the linear structure of the predictor. Thus, like Cox's proposal, it allows a transformation of the predictor in (3.5) that leads to the Cox model for time-dependent covariates

$$\begin{aligned}\lambda(t|x) &= \lambda_0(t) \exp\left\{\beta_0 x + \sum_{j=1}^{m}\beta_j t^{(p_j)}x\right\} \\ &= \lambda_0(t) \exp\left\{\beta_0 x + \sum_{j=1}^{m}\beta_j z_j(t)\right\}\end{aligned} \qquad (3.6)$$

by constructing m time-dependent components $z_j(t) := t^{(p_j)}x$. Fitting model (3.5) therefore reduces to estimating constant effects for the time-dependent covariates $z_j(t)$ in (3.6). Note that $z_j(t)$ belongs to the class of defined (external) covariates described in Chapter 2.

If the hazard depends only on the current covariate value, as in (3.6), the partial likelihood of a Cox model for time-dependent covariates (2.9) can be transformed to the partial likelihood (2.8) of the stratified Cox model, as seen in Section 2.2.4. Failure time then acts as a stratification variable, so that each failure point $t_{(k)}$ defines one stratum, which contains only individuals of the corresponding risk set $R(t_{(k)})$ and a single event. Hence, if the available statistical software package only provides estimation procedures for the stratified Cox model, estimation of (3.5) can also be realised by restructuring the data such that all the risk sets at every failure time-point are matched with the appropriate values for the time-dependent covariate and thus a 'pooled' data set is obtained consisting of $\sum_{k=1}^{K} R(t_{(k)})$ observations. These observations are created in that for each single individual its observation time t_n is split at every preceding failure point $t_{(k)} < t_n$ into 'sub-observations' for which the following data are recorded:

$$
\begin{aligned}
\text{observation time} \ \ &= \text{length of the corresponding subinterval,} \\
\text{state } \delta_{nk} \ \ &= \begin{cases} 1, & \text{for the individual } n = k \text{ that fails at } t_{(k)}, \\ 0, & \text{(censoring) for all others,} \end{cases} \\
\text{index of stratum} \ \ &= k\,, \\
\text{covariate value} \ \ &= z_n(t_{(k)})\,.
\end{aligned}
$$

Finally, a stratified Cox model is estimated by taking the different failure times as a stratification factor.

Determination of the Degree m and the Powers p

In order to fit varying effects using Fractional Polynomials in a flexible way, optimal values of the degree m and of the powers $p_1, \ldots, p_m$ have to be determined in addition. A convenient and practical way is to set an upper limit m_{max} for the degree and prefix a set of possible powers $\mathcal{P}$. The powers are then chosen by selecting the 'best' model out of all models with possible combinations (m-tuples) of $p_j \in \mathcal{P}$ for all $m \leq m_{max}$ due to some goodness-of-fit criterion. Here, it is proposed to minimise the p-value of the likelihood ratio statistic as a non-linear function of fit and complexity. The restriction of the powers to a carefully chosen set $\mathcal{P}$ not only speeds up computation but also assures reasonable interpretations. In practice FPs of maximal degree $m_{max} = 2$ and a set of powers $\mathcal{P} = \{-2, -1, -0.5, 0, 0.5, 1, 2, 3\}$ are usually sufficient. This was also observed and suggested by Royston & Altman (1994), who give examples of the large variety of possible shapes which FPs from this set can take. In particular it contains the conventional polynomials as well as asymptotic functions. Our own experience based on simulations and examples confirm these findings for most applications of dynamic effect modelling.

3.2 Model Choice

3.2.1 Verification of the Dynamic Structure

Before interpreting a dynamic model one is usually interested in verifying the chosen time-varying effect structure, which in particularly implies the assessment of the improvement of the dynamic fit in comparison to the constant fit. This is of special relevance when the focus is on parsimonious modelling and a decision should be taken between the PH-model and the dynamic model, and in turn corresponds to testing the PH-assumption.

A number of graphical and test-based procedures for exploring possible dynamic structures have been proposed in the literature. A broad collection of reviews and references is given by Hess (1994 and 1995). An informal but simple method is to examine whether the constant estimator of the PH-model lies within the standard error bands of dynamic estimation, as obtained by smoothing splines ('SE-method'). Beside such graphical methods, formal goodness of fit tests for time-variation are of particular interest. Following the proposal of Cox this can be done by estimating his extended model (3.3) and then test for its components regarding $H_0 : \beta_1 = 0$ using common test statistics, (e.g. the likelihood-ratio statistic). Omnibus goodness-of-fit tests, which compare the expected and observed frequencies of failures for a given partition of time and covariate space are suggested by Schoenfeld (1980), Moreau, O'Quigley & Mesbah (1985) and Moreau, O'Quigley & Lellouch (1986). Hess (1994) notes that the cubic regression spline approach based on fixed knots allows also for formally testing the PH-assumption. Harrell (1986) recommends to test for correlation between residuals and failure time, where for non-monotonic time-dependencies an 'appropriate' transformation of time has to be used. The PH-test most commonly used is the score test proposed by Grambsch & Therneau (1994), which is based on weighted Schoenfeld residuals. Similar to the Cox proposal they define $\beta(t) = \beta_0 + \beta_1 \varphi(t)$ using some arbitrary transformation of time $\varphi(t)$ and test for $H_0 : \beta_1 = 0$. They showed that this test is equivalent to a generalised least square test, and that its computation only requires the fit of the PH-model under H_0 together with the corresponding Schoenfeld residuals. However, the shape of the suspected departure from time-constancy has to be pre-set. Ng'Andu (1997) gives an overview of various PH-tests with a discussion of their performance in different situations.

The drawback of all these proposals is similar to the problems associated with estimation methods described above. They demand pre-specifications, which are either the suspected functional development over time or a partitioning of the time-axis. Therefore they are in some way only appropriate for testing a certain structure of time-dependency. Even though some of the tests are fairly robust in that the pre-specifications are not required to match the exact time-structure of the effects to identify a PH-violation, the tested alternative does not necessarily provide a suitable model if a departure from the PH-assumption is detected.

An alternative is the method proposed by Gray (1994), who presents a formal test on the dynamic effect structure for fixed knot B-splines with penalty functions. In simulation studies he shows that the test is rather powerful and robust to the number of knots. However, estimation and testing is again based on non-standard methods so that its application is not straightforward. In addition, the fit depends on pre-set smoothing parameters. Also Kauermann & Berger (2002) present a test with a flexible alternative for time-varying effect structures, which is, however, based on bootstrap sampling. As a consequence of all these difficulties, survival analysis within the Cox framework is in practice usually realised in two separate steps: First, the validity of the PH-assumption is investigated based on simple pre-specifications about possible dynamic structures. If the test indicates absence of proportional hazards, an additional, more flexible model is subsequently determined to adequately fit the data.

Naturally, it is desirable to combine these steps in such a way that for verification of the PH-assumption the alternative dynamic Cox model is directly tested against the proportional hazards model. This is attained by the Fractional Polynomial approach, where verification of the PH-assumption $H_0 : \beta(t) = \beta_0$ corresponds to comparing the selected 'optimal' FP model with the PH-model by testing simply $H_0 : \beta_1 = \ldots = \beta_m = 0$. Since this comparison defines a nested hypothesis testing problem, the likelihood ratio test can be applied. Royston & Altman suggest using the χ^2_{2m}-distribution with $2m$ degrees of freedom (df), counting one df for each FP coefficient β_j and one for each selected exponent p_j. The simulation study presented in Section 3.3 supports this suggestion, confirming consistency with the significance level.

3.2.2 Variable Selection

Global Test

Beside testing time variation, it is of interest to examine the global hypothesis $H_0 : \beta(t) = 0$, i.e. to check whether a covariate has a significant influence at all. This is performed considering $2m + 1$ degrees of freedom, $2m$ for the FP and one for constant effect β_0. If the global test is applied after deciding between the FP model and the constant model, it has to be adjusted. In conformity to a stepwise procedure (Kupper et al., 1976), I suggest to correct the test based on Bonferroni's inequality, multiplying the p-value by two. This proves to behave satisfactorily as demonstrated in simulations in Section 3.3.

Confidence Intervals

To visualise the nature of the time-dependency, the resulting FP function may additionally be plotted along with confidence bands. Due to the linearity of the FP-predictor $\eta = \beta_0 x + \sum_{j=1}^{m} \beta_j t^{(p_j)} x$, confidence bands can also be readily computed applying standard estimation techniques:

Using matrix notation, let $\hat{\boldsymbol{\beta}} = (\hat{\beta}_0, \hat{\beta}_1, ..., \hat{\beta}_m)'$ be the regression coefficient estimates and $\mathbf{Z} = (x, t^{(p_1)}x, ..., t^{(p_m)}x)'$ be the matrix of values for the covariate and the time-by-covariate interactions, so that $\hat{\eta} = \hat{\boldsymbol{\beta}}'\mathbf{Z}$. The $(1 - \alpha)$-confidence band for the FP-predictor is then given by

$$\mathrm{CI}_{(1-\alpha)} = \left[\hat{\boldsymbol{\beta}}'\mathbf{Z} \pm (\chi^2_{2m,\alpha/2}\mathbf{Z}'\Sigma\mathbf{Z})^{1/2}\right],$$

where Σ is assumed to be a large-sample covariance matrix for $\hat{\boldsymbol{\beta}}$ and $\chi^2_{2m,\alpha/2}$ is the $(1 - \alpha/2)$- fractile of the χ^2-distribution as described above.

3.2.3   An Algorithm for Multivariate Analysis

The above approach can now be extended to multivariate problems. For the model function itself, this is basically a question of notation. Let $\mathbf{X}$ denote a set of q covariates in arbitrary order. In multivariate dynamic Cox models based on Fractional Polynomials the hazard rate is defined as:

$$
\begin{aligned}
\lambda(t|\mathbf{X}) &= \lambda_0(t) \exp\Big\{ \sum_{i=1}^{q} \beta_i(t) x_i \Big\} \\
&= \lambda_0(t) \exp\Big\{ \sum_{i=1}^{q} [\beta_{i0} x_i + \sum_{j=1}^{m_i} \beta_{ij} t^{(p_{ij})} x_i] \Big\}
\end{aligned}
\tag{3.7}
$$

where FP functions $\beta_i(t) = \phi_{im_i}(t, \mathbf{p}_i)$, $i = 1, \ldots, q$, describe the time interactions of each covariate effect. Model estimation can be realised following the backfitting-strategy proposed by Hastie & Tibshirani (1990), i.e. fitting the coefficients β_{ij} and the exponents p_{ij} iteratively. The backfitting-type procedure starts with fitting the predictor

$$
\eta_1 = \beta_{10} x_1 + \sum_{j=1}^{m_1} \beta_{1j} t^{(p_{1j})} x_1 + \sum_{i=2}^{q} \beta_{i0} x_i
$$

by selecting an optimal FP for covariate x_1, choosing m_1 and p_{1j} $(j = 1, \ldots, m_1)$ as described above for the univariate case. The effects of covariates $x_2, \ldots, x_q$ are kept time-constant in the first step, and only $\beta_{20}, \ldots, \beta_{q0}$ are additionally estimated. Then the likelihood ratio test is applied to assess the gain of fitting time-variation, i.e. the hypothesis $H_0 : \beta_1(t) = \beta_{10}$ is tested as described above. If the time-varying effect of x_1 is distinct, the exponents p_{1j}'s of the first FP are fixed and the predictor

$$
\eta_2 = \beta_{10} x_1 + \sum_{j=1}^{m_1} \beta_{1j} t^{(p_{1j})} x_1 + \beta_{20} x_2 + \sum_{j=1}^{m_2} \beta_{2j} t^{(p_{2j})} x_2 + \sum_{i=3}^{q} \beta_{i0} x_i
$$

is fitted by selecting m_2, p_{2j} for x_2 as above. Note that in this step only the exponents p_{1j} are fixed, while all the coefficients $\beta_{10}, \ldots, \beta_{1m_1}$, $\beta_{20}, \ldots, \beta_{2m_2}$ and $\beta_{30}, \ldots, \beta_{q0}$ are re-estimated. This is continued until η_q is achieved and optimal Fractional Polynomial terms are obtained for all q covariates. In the next iterations the FP functions are similarly updated for each covariate x_i fixing the FPs of the remaining covariates. The algorithm can be stopped when the selected powers $\mathbf{p}_i$, respectively the fit, do not change from one iteration to the next. If a selection of covariates is simultaneously required, the likelihood ratio test of $H_0 : \beta_i(t) = 0$, $i = 1, \ldots, q$, can be investigated additionally in each iteration. A covariate can then be omitted in one iteration if it does not provide sufficient improvement to the fit. It can be re-entered and re-tested in later iterations.

The order of the covariates is generally irrelevant if the covariates are independent. For dependent covariates however, the results of the algorithm may depend, like any iterative procedure of this type, on the order of selection. To assure reproducibility a general suggestion is to fix the order with respect to the p-values of a full PH-model. Still, in this case further analysis of the finally selected FP model – especially of its dynamic structure – has to be performed on a substance matter basis, which may possibly require a reselection of the model. This is however, a common problem in multivariate analysis not specific to FP modelling and thus not further discussed here.

3.3 A Simulation Study

The performance of the Fractional Polynomial approach shall now be analysed in a simulation study, with special focus on its capability to detect violations of the PH-assumption, its power in covariate selection when the effects vary over time and its dependency on sample size, censoring rate and event probability.

First, 1000 samples of failure-time data are generated from five different scenarios. All samples consist of two groups, a 'baseline group' with $x = 0$ and a 'risk group' with $x = 1$, each of which have 100 observations. To allow a precise simulation of temporal effect structures the data are generated from a logistic setting, where the 'baseline hazard of failure' at each time-point is set to $\lambda_0 = e^{-4}/(1 + e^{-4}) = 0.018$. Dynamic effects for x are simulated using five different functions (compare Figure 3.1 (a)):

(1) A constant model with proportional hazards: $\beta(t) = 1$, to analyse consistency of the PH-test $H_0 : \beta(t) = \beta$.

(2) A linear time-dependency: $\beta(t) = -0.02t + 1$, describing an effect, that decreases over time.

(3) A steep quadratic time-dependency: $\beta(t) = -1 + 0.08t - 0.0008t^2$, describing an effect which rises in the beginning and decreases later as certain treatment effects do (see e.g. Stablein et al., 1981).

(4) A flat quadratic time-dependency: $\beta(t) = -0.5 + 0.04t - 0.0004t^2$ to examine the sensitivity of the FP-approach.

(5) A step function: $\beta(t) = 1.5$ for $t \leq 24$ and $\beta(t) = 0$ for $t > 24$, describing an effect which abruptly disappears.

Additionally, a permanent probability P_c for (right) censoring with value 0.005 is assumed for all groups. This produces an average censoring rate of about 25 % for the baseline group. These settings generate survival data with few ties and rather long survival times, so that there are no objections against employing a Cox model. Moreover the structure of the data is very similar to what is known from clinical studies.

For the simulated data sets (1)-(5) the optimal FP functions are determined and tested for dynamic effect structures as described in Subsection 3.1.2. The empirical type I error (α_{PH_emp}) for a nominal significance level of $\alpha_{PH} = 0.05$ is obtained as the number of p-values less than α_{PH}, divided by the simulation size. For comparison the results obtained from the following, commonly used proposals are recorded as well:

- A test based on the extension (3.3) proposed by Cox (1972) using a) a linear transformation $\varphi(t) = t$, b) a logarithmic transformation $\varphi(t) = \log(t)$ and c) a quadratic transformation $\varphi(t) = t^2$.

- The residual score test of Grambsch & Therneau (1994) using d) the Kaplan-Meier transformation, e) the linear transformation, f) the logarithmic transformation and g) the quadratic transformation.

- The informal 'SE-method' using the smoothing spline estimators (with 4 degrees of freedom) from Hastie & Tibshirani (1993), checking if the estimated effect β_0 of the constant Cox PH-model lies outside the 2· standard errors bands, h) for any event point and i) for more than 10 % of the event points.

Figure 3.1 pictures the proportion of cases where the hypothesis of constant effects, i.e. the PH-assumption, is rejected. Figure 3.1(b) illustrates the consistency of the different tests when the PH-assumption holds, i.e. the empirical type I error. It shows that the FP approach is slightly too liberal and detects time-variation in 6.2 % of the cases. A closer look at these 'type I error'-cases shows that 75 % (47) of them were fitted by a FP of degree $m = 1$ only, and half of these (23) using asymptotic functions with powers $p < 0$ which approach a constant value quickly. In contrast, the test of Grambsch & Therneau is too stringent (1 % - 3 %), while

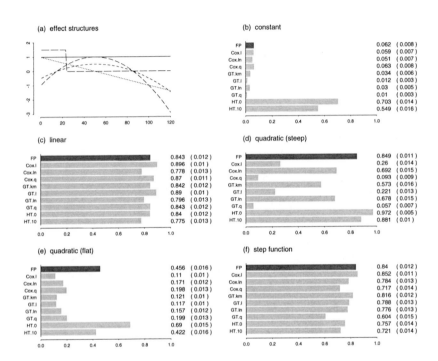

Figure 3.1: Simulated effect structures and the corresponding powers resp. error rates of different PH-tests at a significance level of $\alpha = 0.05$ (SE in brackets). FP: Fractional Polynomial, Cox.l: Cox's test with linear transformation, Cox.ln: Cox's test with logarithmic transformation, Cox.q: Cox's test with quadratic transformation, GT.km: Test of Grambsch & Therneau with KM transformation, GT.l: Test of Grambsch & Therneau with linear transformation, GT.ln: Test of Grambsch & Therneau with logarithmic transformation, GT.q: Test of Grambsch & Therneau with quadratic transformation, HT.0: More than 0 % outside 2·SE bands of the smoothing spline, HT.10: More than 10 % outside 2·SE bands of the smoothing spline.

the test based on Cox's proposal is closest to the nominal α-level. The heuristic SE-method clearly decides upon temporal structures far too often and even its relaxation to a 10 % -limit causes an error of 54.9 %. This is not surprising, as it is well known that point-wise confidence bands do not allow for a global interpretation. Still, the degree of violation of the significance level is remarkable.

Figures 3.1 (c)-(f) present the power of the different tests for the four dynamic settings. In the linear setting (Figure 3.1(c)) all tests detect the PH-violation about equally well (77.8 % - 89.6 %). In contrast, in the simulations from the steep quadratic function (Figure 3.1(d)) with a mode at time point 50 , the power of the various tests differ noticeably. With the FP approach, significant time-variation is detected in 84.9 % of the situations, while the test based on Cox's proposal and the residual score-test of Grambsch & Therneau even in the best case only find significant PH-violation in about 2/3 of the samples. Note that the reason for the extremely low power of these tests when using the quadratic transformation $\varphi(t) = t^2$ (9.3 % and 5.7 %) is due to the missing linear term. In both tests the verified time-transformation is expressed in a single term. Therefore, to achieve a better power the time-transformation must be prespecified exactly by $\varphi(t) = 100t - t^2$ shifting the mode from 0 to 50.

The results of the simulations based on a flat quadratic temporal function (Figure 3.1(e)) are of particular interest. Here the FP approach with its flexible functions is clearly superior (45.6 %) to the test of Cox and the residual score-test of Grambsch & Therneau, which is hardly able to detect the time-variation with its predefined time transformations (power: 11.7 % - 19.9 %). The interpretation of the SE-method for the dynamic settings is inappropriate due to lack of consistency. Finally, simulations based on the step function (Figure 3.1(f)) illustrate the performance of the PH-test when the underlying dynamic structure is clearly different to those functions constructed from the power-set $\mathcal{P}$. It appears that the FP approach provides high power in detecting the time-variation in this situation, too. Although the step function might be seen as artificial in its abruptness, in practice dynamic structures could be relevant which show a strong effect, which disappears after some time. For an accurate description of those effects it might be necessary to extend the set of FP functions. When bell-shaped curves of the type $\exp\{-t^2/(\tau \cdot \bar{t}_{med})\}$, $\tau = 0.5, 1, 2$, to model the samples from the step function are included, they are selected in 55 % of the cases and raise the power of the PH-test to 88.4 %. For more flexibility, $\tau \cdot \bar{t}_{med}$ could be replaced by an additional coefficient, which would then have to be determined by a grid search (Sauerbrei & Royston, 1999), or by iterative optimisation (Royston, 2000a).

To complete the picture on PH-testing, I additionally study the performance of the following alternative tests based on dynamic fits:

- The 'informal' deviance-based test comparing the fit of the smoothing spline (with 4 df) and the PH-fit using a χ^2-distribution (Hastie & Tibshirani, 1990, p. 155).

- The test proposed by Hess (1994) based on cubic regression splines with three and five knots.

- The piecewise test, fitting constant effects in prespecified segments of time using four knots, (Moreau et al., 1985).

Table 3.1: Power resp. error rates for H_0 : $\beta(t) = \beta$ at level $\alpha_{total} = 0.05$, for PH-tests based on dynamic fits for different settings. (SE in brackets)

setting	FP	smoothing spline	Hess (3 knots)	Hess (5 knots)	piecewise
Const. model	0.062	0.016 (0.008)	0.052 (0.014)	0.060 (0.015)	0.024 (0.010)
Linear	0.843	0.728 (0.028)	0.832 (0.024)	0.772 (0.027)	0.716 (0.029)
Steep quadratic	0.849	0.852 (0.022)	0.916 (0.018)	0.876 (0.021)	0.652 (0.030)
Flat quadratic	0.456	0.452 (0.031)	0.592 (0.031)	0.488 (0.032)	0.380 (0.031)
Step function	0.840	0.648 (0.030)	0.872 (0.021)	0.820 (0.024)	0.772 (0.027)

250 replications are generated from the settings described above. The results are shown in Table 3.1 and compared to the results for the Fractional Polynomials from above. The deviance-test for the smoothing splines yields a far too small empirical type I error ($\alpha_{emp} = 0.016$) in the constant setting. This reflects the fact that in the case of smoothing splines, the deviance is not even asymptotically χ^2 distributed, (Hastie & Tibshirani, 1990). Yet, in complex dynamic settings, such as the quadratic functions, this test turns out to be rather powerful. In the linear setting though its power is small. This can be traced back to the penalty used for smoothing, i.e. $\frac{1}{2}\tau \int \beta''(t)^2 \partial t$. Here, the smoothing parameter τ controls the complexity of the model and directly affects the number of degrees of freedom of the model which is used for testing. For smaller τ the degrees of freedom increase and

the fit becomes rougher. For $\tau \to \infty$ the resulting model tends to select a linear effect $\beta(t) = \beta \cdot t$. Any model resulting from a smaller τ will be more complex using more degrees of freedom, which is superfluous in the linear situation. Note that the test is also less powerful in identifying the step function. For similar reasons the test suggested by Hess (1994) based on cubic regression splines is also less powerful in the linear setting, since for decreasing number of knots the regression spline converges to a linear function of time. Even though this test appears to behave satisfactorily in the other situations, the simulations indicate its dependency on the choice of the number of knots: For our settings with 200 observations, regression splines with five knots seem to be too complex reducing the power of the test. The test based on a piecewise constant fit using four knots is overall less powerful in identifying smooth time-variations .

It should be noted that all these three methods produce piecewise fits of the dynamic structure on a specified grid. While regression splines can be seen as a smooth extension of piecewise constant fitting, natural smoothing splines based on optimisation of the penalised partial likelihood (3.2) result in a spline function with knots at each failure time-point, (see Section 3.1). Altogether, the tests based on these fits suffer from their dependence on the chosen smoothing parameter or the number and location of knots. In applications a large number of degrees of freedom or knots is often used to assure a flexible fit. This however, adversely affects the power of these tests when the true dynamic structure is in fact simple or even linear.

Table 3.2: Power resp. error rates for $H_0 : \beta(t) = 0$ at level $\alpha_{total} = 0.05$, for different settings for sample size, censoring rate and event probability. (SE in brackets)

setting	adj. FP model	unadj. FP model	Cox PH-model
Null model I $(n = 200)$	0.043 (0.006)	0.081 (0.009)	0.044 (0.006)
$(n = 200)$			
Linear	0.935 (0.008)	0.967 (0.006)	0.706 (0.014)
Steep quadratic	0.873 (0.011)	0.914 (0.009)	0.543 (0.016)
Flat quadratic	0.374 (0.015)	0.478 (0.016)	0.109 (0.010)
Step function	0.999 (≈ 0)	0.999 (≈ 0)	0.999 (≈ 0)
Null model I $(n = 400)$	0.048 (0.010)	0.076 (0.012)	0.040 (0.009)
Null model I $(n = 800)$	0.038 (0.009)	0.082 (0.012)	0.048 (0.010)
Null model II $(n = 200)$	0.042 (0.009)	0.076 (0.012)	0.042 (0.009)
Null model III $(n = 200)$	0.046 (0.009)	0.080 (0.012)	0.048 (0.010)

The next objective is to examine the performance of the global test on $H_0 : \beta(t) = 0$ for the four dynamic scenarios from above and for a null model with no risk-effect (Null model I) where the baseline settings from above are used to simulate 1000 replications. Again a nominal significance level of $\alpha_{total} = 0.05$ is considered. Since a prior choice between the FP model and the constant model was made, the global p-values should be adjusted according to Bonferroni, multiplying them by two (see Section 3.2.2). In Table 3.2 the results of this test with ('adj. FP model') and without ('unadj. FP model') application of Bonferroni's correction are given and compared to the results of the Cox PH-model. The simulations show that the adjusted FP approach tends to be slightly conservative. Still, the accuracy seems acceptable, especially when seen in comparison to the empirical type I error of the Cox model. The need of correcting the p-value is emphasised by the results of the unadjusted test (unadj. FP model), where the value of α_{emp} is inflated. Furthermore, the FP approach yields high power in detecting effects that vary over time in the dynamic settings, while the Cox PH-model with its constant coefficient naturally often misses to detect these effects. For comparison, I additionally determined the global test for the smoothing spline estimates, which are also based on the deviance and the χ^2-distribution. Again the test is rather conservative with $\alpha_{emp} = 0.023$ ($SE = 0.005$), but yields slightly better power in the quadratic settings (steep: 0.892, SE = 0.009; flat: 0.401, SE = 0.015). In the linear setting it is slightly less successful in detecting the linear effects (0.886, SE = 0.01).

To study the influence of sample size, censoring rate and event probability, the simulation settings of the null model are varied. First, the sample size is raised to 200 and 400 observations per group, obtaining $n = 400$ and 800, respectively. In Null model II the censoring probability is increased to $P_c = 0.02$, which produces data with an average censoring rate of about 50 %. Finally, in Null model III, I decrease the baseline hazard to $\lambda_0 = e^{-5.5}/(1 + e^{-5.5})$ and set $P_c = 0.002$, producing longer survival times and an average censoring rate of about 35 %. For each setting I generate 500 replications and perform the global test. The results of these simulations, given in the lower part of Table 3.2, are similar to above and do not show any dependency on sample size, censoring rate or event rate.

Summarising the simulation results, it is seen, that Fractional Polynomials provide a reliable tool to investigate violations of the PH-assumption. They are flexible enough to yield high power in different situations of time-variation while the test is consistent with the significance level. The comparison of the FP approach to other methods by the simulation study is focused primarily on those tests which are either easy to apply, as the test based on Cox's proposal, readily available, as the S-Plus function determining the residual score test of Grambsch & Therneau, or widely used such as it is the case for the exploration of error bands using natural smoothing splines (Figure 3.1). It shows that the test based on Cox's proposal and the score-test of Grambsch & Therneau, which use predefined time-transformations, are not flexible enough to properly identify time-variation of complex structure. The graphical method for checking the PH-assumption based on a comparison of the constant Cox estimator and point-wise standard error bands of natural smoothing splines clearly violates consistency, underlining the fact that it is not suitable for global assessments. The tests based on comparison of dynamic fits with the PH-model, such as smoothing splines, regression splines and piecewise constants, are rather powerful in complex settings. However, these tests are less powerful in detecting linear dependencies, which are common in survival analysis. Moreover they severely depend on the choice of the smoothing parameter, respectively the number and location of knots. To examine the performance of the FP approach in detecting the overall impact of a covariate, the global hypothesis $H_0 : \beta(t) = 0$ was tested by employing an adjusted likelihood ratio test. It provides powerful results that are robust against changes in sample size, censoring rate and event rate, (see Table 3.2).

3.4   Time-Variation in Gastric Cancer Prognosis

For a study of prognostic factors and risk-group stratification in gastric carcinoma at the university hospital of the Technische Universität München, the survival of gastric cancer patients was followed up after complete resection of the tumour. One major point of interest of this study was to investigate the prognostic impact of two new tumour-biological factors, the urokinase-type plasminogen activator uPA and its type-1 inhibitor PAI-1 (Nekarda et al., 1994). UPA and PAI-1, both assessed in extracts of cancer tissue at surgery, belong to the plasminogen activator system, which has been reckoned to play an important role in tumour cell migration. Clinically it is of great interest to ascertain whether there is time-variation in the effect of these factors.

Table 3.3: Prognostic factors analysed in the gastric cancer study

factor	range	coding	interpretation
AGE	28-90	0: $\leq$ 65 1: $>$ 65	Age at surgery
pN.RATIO	0-97	0: $<$ 20 % 1: $\geq$ 20 %	Percentage of positive lymph nodes
pT.JAP	1-7	0: $\leq$ 4 (pT1 - pT2a) 1: $>$ 4 (pT2b - pT4)	local tumour invasion (Japanese staging system)
pM	yes/no	0: no 1: yes	distant metastasis
uPA	0.02-20.57	0: $\leq$ 1.2 ng/mg 1: $>$ 1.2 ng/mg	urokinase-type Plasminogen Activator
PAI-1	0.02-264.62	0: $<$ 4.13 ng/mg 1: $\geq$ 4.13	Plasminogen Activator Inhibitor Type 1

3.4.1 The Data

295 gastric cancer patients were enrolled between 1987 and 1996, 108 of whom died during follow up. Time to death, measured in months, is used as failure time. The median follow-up time is 41 months. In addition to the two proteases uPA and PAI-1, established clinical factors such as age and TNM-classification are included in the analysis, where TNM consists of local tumour invasion, lymph node involvement and distant metastasis. All factors are regarded in binary coding. Table 3.3 gives a short description of the factors. Local tumour invasion (pT.JAP) was assessed following the Japanese staging system for gastric cancer, which classifies the depth of invasion. It is binary coded, separating invasion of Mukosa to Subserosa (pT1-pT2a) from perigastric fat tissue to neighbouring organ (pT2b-pT4). Lymph node involvement (pN.RATIO) is quantified as the ratio of invaded to removed lymph nodes and binary coded at a level of 20 % (Nekarda et al., 1994; Siewert et al., 1998). Age at surgery was dichotomised at 65, which corresponds to the median.

For the new factors uPA and PAI-1, measured in ng/mg Protein, cutpoints used for binary coding were selected by optimisation of the log-rank-statistics. Both correspond approximately to the lower quartiles.

3.4.2 Results

Univariate Cox PH-Analysis

Table 3.4 summarises the results of the univariate Cox PH-models for each factor, giving the factor's effect β, the relative risk **RR** and its p-value from the likelihood ratio test. In addition the p-value from the residual score test of Grambsch & Therneau was calculated using a logarithmic transformation of time.

Table 3.4: Results of univariate Cox PH-models

| factor | Cox PH-fit | | | residual score test |
| | | | $H_0 : \beta = 0$ | $H_0 : \beta(t) = \beta$ |
	β	**RR**	p-value	p-value
AGE	0.18	1.20	0.36	0.002
pN.RATIO	1.88	6.55	<0.001	0.223
pT.JAP	1.55	4.71	<0.001	0.795
pM	1.37	3.94	<0.001	0.880
uPA	0.81	2.25	0.002	0.885
PAI-1	1.25	3.49	<0.001	0.975

All factors except age show a statistically significant impact on survival at a level of $\alpha = 0.05$. The residual score test however, identifies significant time-variation for the effect of AGE indicating a violation of the PH-assumption. The crossing Kaplan-Meier curves for the groups with age ≤ 65 and age > 65 (Figure 3.2(a)) support this presumption.

Multivariate Cox PH-Analysis

Table 3.5 shows the results of the Cox PH-model comprising all six covariates. The percentage of positive lymph nodes and local tumour invasion turn out to be the strongest prognostic factors, increasing the risk of death by a factor of 4.47 and 2.72, respectively. In addition, the proteolytic factor PAI- 1 shows statistically significant impact on survival. In this model however, beside age also uPA and distant metastases are no longer significant. The test based on scaled Schoenfeld residuals with a logarithmic transformation again detects a dynamic structure for the effect of AGE.

Table 3.5: Results of a multivariate Cox PH model

factor	Cox PH-fit			residual score test
			$H_0 : \beta_i = 0$	$H_0 : \beta_i(t) = \beta_i$
	β_i	**RR**	p-value	p-value
AGE	0.37	1.45	0.065	<0.001
pN.RATIO	1.50	4.47	<0.001	0.178
pT.JAP	1.0	2.72	<0.001	0.134
pM	0.23	1.26	0.35	0.458
uPA	0.18	1.20	0.52	0.476
PAI-1	0.87	2.39	0.007	0.591

To further explore the temporal structures the proposed FP-method is employed, restricting the degree of the polynomials to $m \leq 2$ and fixing the set of relevant powers to $\mathcal{P} = \{-2, -1, -0.5, 0, 0.5, 1, 2, 3\}$, as stated in Section 3.1.2.

Univariate FP-Analysis

The univariate FP model for AGE confirms time-variation. Its significant effect decreases over time ($p_{PH} = 0.005$, $p_{total} = 0.02$), following the FP function $\beta(t) = -1.28 + 4.94 \cdot t^{-0.5}$ which is shown in Figure 3.2(b). Shortly after surgery, older patients (65 years and older) have a higher mortality rate. This difference declines with time and after about two years of follow-up the younger patients appear to have a higher risk. For all other factors the univariate FP approach re-

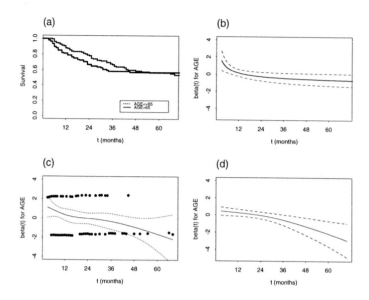

Figure 3.2: (a) Kaplan-Meier survival curves of the groups with age $\leq$ 65 and age > 65. (b) FP function for AGE with ± 2 standard error bands. (c) Raw and spline-smoothed scaled Schoenfeld residuals for AGE, with ± 2 standard errors. (d) The natural smoothing spline estimator of $\beta(t)$ for AGE with 95 %-confidence limits.

sults in time-constant effects. Figure 3.2(c) gives the scaled Schoenfeld residuals for the univariate model with AGE, together with a scatter-plot spline-smoother and ± 2 standard error bands. Again the dynamic structure can be seen. For further comparison I additionally determine the natural smoothing spline estimator (with 4 df) for $\beta(t)$, given in Figure 3.2(d), which shows an even sharper decline in the end.

Multivariate FP-Analysis

To fit a multivariate dynamic Cox model the backfitting-type algorithm of Section 3.2.3 is used, which enables a data-driven decision on time-variation for each factor (Table 3.6). Beside age, the effect of distant metastases also shows a significant change over time. Figure 3.3 pictures their FP estimators. It again indicates

Table 3.6: Results of the multivariate FP model

factor	$\beta_i(t)$	m	$H_0 : \beta_i(t) = \beta_i$ p-value	$H_0 : \beta_i(t) = 0$ p-value
AGE	$\beta_1(t) = 0.98 - 0.001 \cdot t^2$	1	<0.001	<0.001
pN.RATIO	$\beta_2(t) = 1.49$	0	—	<0.001
pT.JAP	$\beta_3(t) = 1.08$	0	—	<0.001
pM	$\beta_4(t) = 0.12 + 2.65 \cdot t^{-2}$	1	0.033	0.089
uPA	$\beta_5(t) = 0.24$	0	—	0.779
PAI-1	$\beta_6(t) = 0.90$	0	—	0.003

a distinct decrease of the effect of age over time. The effect of distant metastases declines as well, following an asymptotic course and staying constant after about one year. For comparison also the natural smoothing spline estimators (with 4 degrees of freedom) from a multivariate analysis are plotted. For the effect of age they show the same dynamic structure as the FP function. Apparent disagreement can be observed between the FP function and the spline estimator for the effect of distant metastasis, where the spline estimator shows a sharply decreasing tail. This is due to the small number of events towards the end of follow-up in the shrunken risk group with distant metastases 'pM = 1', rather than an increased event rate for 'pM = 0'. Smoothing spline estimators generally tend to be influenced more by late events and often show a disproportionate drift in the tail. Due to their fixed smoothing parameter, this may cause too flat courses in the beginning. When all remaining data is censored after 60 months the smoothing spline obtains a flat tail and a slightly steeper front. The FP-model on the other hand is less sensitive to artefacts caused by long survivors.

The adjusted test on global effects shows that in this multivariate analysis uPA doesn't provide additional information on prognosis, which is in agreement with the results of Table 3.5. Also pM is no longer significant when considering a significance level of $\alpha_{total} = 0.05$. This is due to the correlation of the TNM-factors. Gastric cancer patients with distant metastases usually have either a larger number of involved lymph nodes, a deep tumour invasion or both. Therefore, if pT.Jap and pN.RATIO are included into the model, the additional impact of pM is reduced. If pM and uPA are omitted, the fit of the remaining factors basically stays the same, choosing exactly the same FP-functions and only very slightly adapting the coefficients. Also reordering of the factors in the multivariate algorithm has no influence on the finally selected model.

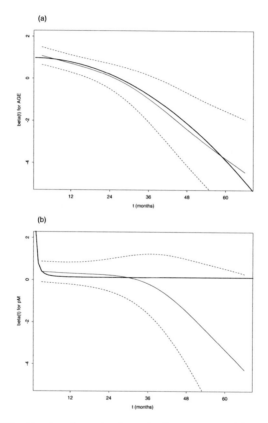

Figure 3.3: FP estimators from the dynamic Cox model for (a) AGE and (b) pM (thick line). For comparison the smoothing spline estimators of $\beta(t)$ are plotted (——), with 95 %-confidence limits (- - - -).

3.4.3 Summary

The example of gastric cancer data clearly illustrates the general importance of correctly specifying existing dynamic effect-structures. The prognostic impact of the factor age on survival was not identifiable in the PH-model. In the multivariate dynamic model next to lymph node involvement, tumour invasion and age, the new proteolytic factor PAI-1 shows additional prognostic information.

In the analysis all factors were included in 0/1 coding, to assure a better interpretation of the outcome and to improve model robustness by reducing susceptibility to selection effects at late follow-up or to outliers. Certainly, binary coding of the covariates means a loss in diversity and hence information may be lost when differentiating the tumours, which is especially visible for the TNM-factors (i.e. local tumour invasion, lymph node involvement and distant metastasis). Besides, if binary coding is based on optimised cut-points, such as for PAI-1 and uPA, the p-values and effect-estimators from these factors might be too optimistic. In any case, the selected model needs to be validated before being applied in clinical practice.

3.5 Discussion

In this chapter Fractional Polynomials were introduced in order to detect and model dynamic effects in survival data within the Cox model framework. The substantial benefit of this approach is that FP functions are linear in the regression coefficients and the estimation problem reduces to fitting a Cox PH-model with time-dependent covariates. Therefore, model estimation is straightforward, following the strategy described in Section 3.1.2, where standard methods of inference are used, which are readily available in most statistical packages. In addition, the FP approach allows for testing the time-varying effects using standard methods such as the likelihood ratio test. Their flexible construction ensures omnibus power, as is shown in the simulation study.

The comparison with other methods reveals the advantages of the Fractional Polynomials approach. Tests that demand a prespecification of the suspected time-variation do not provide omnibus power and are especially weak when the true dynamic effect structure is complex. In contrary, tests based on a flexible, piecewise fit are less powerful in detecting linear or quasi-linear time-variations. However, in applications, time-varying effects that develop in a linear way are especially common, e.g. when the impact of a covariate decreases over time. Decisions based on a graphical comparison of the constant effect and the fit of a smoothing spline ('SE-method') appear to be generally inappropriate.

Beside the exploration of dynamic effect structures, also variable selection under consideration of varying effects is straightforward using Fractional Polynomials. The simulation studies on the global hypothesis $H_0 : \beta(t) = 0$ showed that the adjusted test yields satisfying results.

The application to gastric cancer data showed that Fractional Polynomials provide a flexible fit and are less sensitive to artefacts, often caused by long term survivors. The set of powers $\mathcal{P}$ that was used to fit FP's (see Section 3.1.2) is tentative and might be adapted in different situations. There are, for example, cases, as showed in the simulations from a step function, when one would prefer to enlarge the set by other transformations of t, or to substitute some of them. In some applications, there might be even some prior knowledge about the dynamic structure, which can be taken into account in the model building process. Nevertheless, the set $\mathcal{P} = \{-2, -1, -0.5, 0, 0.5, 1, 2, 3\}$ already covers a wide range of possible shapes and is usually sufficient.

In summary, Fractional Polynomials offer a useful tool for analysing survival data when proportionality of the hazards is in doubt but no pre-information about the dynamic effect structure is given. They describe the relationship between the effect of the factor and time with simple functions, which are stable and easily communicated. By allowing for variable selection and verification of the dynamic structure, the multivariate analysis results in a parsimonious model where only those effects are modelled in a dynamic way, for which evidence for time-variation is given in the data. Still, although Fractional Polynomials provide a flexible fit, their global (non-local) definition suffers from the same restrictions as other non-local definitions of smooth function estimation. If a fine drawing of the variation is essential, the FP approach could be used in a hybrid-like algorithm to select the important covariates and to decide on their effect structure, while afterwards further semi-parametric modelling methods could be applied. Royston (2000b) recently presented a strategy to verify, whether the global, parametric fit of an additive model based on FP's misses important information. The test he proposes can in principal be transferred to our dynamic effect problem to investigate if the FPs miss out some of the dynamic effect structures.

Chapter 4

The Bayesian Dynamic Logit Model for Survival Data

In the previous chapter an approach has been introduced for modelling continuous survival data by a dynamic Cox model, where the effects of the prognostic factors are allowed to change over time. If survival time is discrete the Cox model may yield inaccurate results and, as described in Chapter 2, discrete models such as the logit model are more appropriate. Thereby Bayesian statistics offers an attractive alternative to maximum likelihood based statistics. Bayesian posterior estimation based on Markov Chain Monte Carlo simulation is universally applicable and provides a convenient framework to flexibly model complex, high-dimensional effect structures. Moreover, the Bayesian approach does not require approximate normality assumptions for estimates, so that these methods are also useful for modelling data of moderate sample sizes.

In this chapter a Bayesian approach to dynamic modelling of discrete survival time data is considered. To fit time-varying effects a model definition in form of a state-space model is employed. It describes the development of the effects over time in a separate model equation constituting a hierarchical model structure. Posterior estimation of the different model paramteters is based on MCMC sampling using a hybrid algorithm.

While with the introduction of Markov Chain Monte Carlo methods Bayesian modelling became more and more popular, Bayesian model diagnosis still requires substantial research. Therefore a particular focus of this chapter is on methods for Bayesian model comparison, which can be used to decide if an effect varies over time or not. Different model selection criteria are investigated, which originate from different theoretical considerations. These include the classical Bayes factor and the Bayesian deviance as well as other proposals that are deduced from these two concepts.

The chapter is organised as follows: Section 4.1 gives a brief introduction to the Bayesian theory of statistics and Bayesian inference, outlining the differences to maximum likelihood theory used in Chapter 3. In particular two fundamental theorems of Bayesian statistics are described, which are used in the following sections. In Section 4.2 a hierarchical logit model for modelling discrete survival time data is defined. Its semi-dynamic predictor includes time-constant and time-varying effects. Smoothing of the dynamic effect functions is performed by a second order random walk, which can bee seen as a discrete analogue to smoothing splines. This modelling approach allows for a joint determination of the smoothing parameter and of the posterior estimates of the dynamic effects via Markov-Chain Monte-Carlo sampling, which is described in Section 4.3. In Section 4.4 different criteria for model selection are investigated with resrpect to their performance in comparing Bayesian non-nested hierarchical models of arbitrary dimension. Since the dynamic logit model for survival analysis is rather complex and posterior determination demands for a considerable computational effort, a special focus is laid on computational aspects of the criteria. In Section 4.5 a comparative simulation study is presented, analysing the performance of the different criteria in order to discriminate between Bayesian survival models with dynamic predictors. The criteria are then employed for model selection in an application in Section 4.6, where the impact of various determinants on infant mortality in Zambia is investigated. It is thereby suspected, that the effects of some of the factors change over the age of the child. Hence the exploration of possible dynamic effect structures is of particular interest. This includes criteria-lead decisions on which of the covariates demand for a dynamic effect. The different Bayesian model criteria are finally discussed in more detail in Chapter 5, where both, the differences of the underlying concepts as well as the relations between them are explored.

4.1 Preamble to Bayesian Statistics

The Bayesian approach to statistics fundamentally differs from maximum likelihood statistics as it formally takes prior information into account which had been available *before* any data was observed. This prior information is then converted into posterior information using the observed data, by employing the Bayes theorem, established by Thomas Bayes (1701-1761). Formally this means that in a Bayesian statistical model both, observables and parameters, are considered to be random quantities. Hence, if data y have been observed and θ denotes the entire set of parameters of a model $\mathcal{M}$ formal Bayesian inference requires the consideration of the joint distribution of all random quantities, that is $P(y, \theta)$, which is given by

$$P(y, \theta) = P(y|\theta) \cdot P(\theta). \tag{4.1}$$

The joint distribution is composed of two parts: $P(y|\theta)$ is the conditional probability of the data y given the parameters θ, which is the *likelihood of the data* and therefore will be termed $l(y|\theta)$ or shortly l, as before. This is supplemented by the probability of the parameters $P(\theta)$ which carries all *prior information* on the non-observables. For a clear distinction from other probabilities, it will be denoted by $\pi(\theta)$. For ease of notation the model index $\mathcal{M}$ is omitted throughout this section.

To make inference about the parameters θ conditional on the observed data y, interest lies in the conditional distribution $P(\theta|y)$, that is the *posterior distribution* of the parameters. The relation between the prior information, the likelihood and the posterior distribution of the parameters is defined by Bayes theorem.

The Bayes Theorem

The Bayes theorem defines the posterior distribution of the model parameters θ given the data y by

$$P(\theta|y) = \frac{l(y|\theta) \cdot \pi(\theta)}{\int l(y|\theta)\pi(\theta)d\theta}. \tag{4.2}$$

The integral in the denominator of (4.2) is the marginal probability of the data, $\int_\Theta l(y|\theta)\pi(\theta)d\theta = P(y)$. Expressing (4.2) in words, it is

$$\text{Posterior} = \frac{\text{Likelihood} \cdot \text{Prior}}{\text{Marginal Probability}}.$$

Since the marginal probability $P(y)$ does not depend on the parameters θ, the posterior distribution is proportional to the likelihood times the prior, which is with (4.1)

$$P(\theta|y) \propto l(y|\theta) \cdot \pi(\theta) = P(y, \theta). \qquad (4.3)$$

This is known as the 'principle of inverse probability'.

The posterior distribution $P(\theta|y)$ is used to draw inference about the model parameter, e.g. point estimates may be obtained by the posterior mean or the posterior median together with precision measures such as the posterior standard deviation or credibility regions. Generally all these characteristic can be expressed in terms of posterior expectations of functions of θ, which are given by

$$E_{\theta|y}[f(\theta)|y] = \int_{\Theta} f(\theta) P(\theta|y) = \frac{\int f(\theta) l(y|\theta) \pi(\theta) d\theta}{\int l(y|\theta) \pi(\theta) d\theta} . \qquad (4.4)$$

The Bayesian Central Limit Theorem

An essential property of the posterior distribution is its *prior robustness* for large samples. It seems natural that in the presence of precise information provided by a large or highly informative data set, the prior should have little or no effect on the posterior information. This is expressed by the *Bayesian central limit theorem*, which says that for a large sample and under commonly satisfied assumptions, the posterior distribution of the parameters θ is approximately normal. The Bayesian central limit theorem can be expressed in different forms, yielding different levels of accuracy of the approximation. The version most commonly used is given by

$$\theta|y \sim N(\hat{\theta}, [-L''_{\hat{\theta}}]^{-1}) \qquad (4.5)$$

where the $\hat{\theta}$ is the maximum likelihood estimate for θ, and $[-L''_{\hat{\theta}}]^{-1}$ is the inverse observed Fisher information evaluated at the maximum likelihood estimate, with

$$L''_{\hat{\theta}} = \frac{\partial^2}{\partial\theta\partial\theta'} \log l(y|\theta) \bigg|_{\hat{\theta}} .$$

Property (4.5) will play an important role when arguing for different model criteria in Section 4.4. A heuristic proof of this approximation via Taylor expansion about $\hat{\theta}$ is found in Berger (1985, ch. 4) together with more details on the other approximations and a list of further references.

4.2   A Bayesian Logit Model with Dynamic Effects

To flexibly model discrete survival data with time-varying effects a dynamic logit model seem suitable, as described in Section 2.3. The Bayesian modelling approach, described here, provides a convenient solution for modelling dynamic effect structures within this framework by including a discrete penalty term. The model definition has a hierarchical form, where different prior specifications are introduced on separate levels. The first level is formed by the *observation model*, which relates the covariates to the outcome conditional on the model parameters. To incorporate dynamic effect structures, the observation model is supplemented by a stochastic transition model, i.e. the *parameter model*, which describes the development of the effects over time. The definition of the parameter model again conditions on hyperparameters. These may either be fixed to a pre-specified value or may be stochastic and follow a prior model themselves which is defined on a further level. Altogether this yields a model of state-space structure as described in Fahrmeir & Tutz (2001, ch. 8). Such hierarchical model definitions are especially attractive as they can be easily extended by modifying the observation model and adding additional prior specifications on separate levels. In this way, semi-dynamic models can be defined, where only part of the parameters have time-varying effects. Priors of the constant effects are specified separately in an additional level, e.g. by normal distributions. In the same way the model could also be extended to include further effect structures such as random or spatial effects.

The Observation Model

In the survival context, the observation model describes the relation between the covariates and the hazard function for failure. For discrete survival time the dynamic logit model as introduced in Chapter 2, is employed. For convenience it is briefly summarised again.

Consider discrete survival time $t = 1, \ldots, K$, where for interval censored survival data, t shall denote the interval $[a_{t-1}, a_t)$. Otherwise t may represent time units such as days, months or years. To simplify the notation it is assumed that the intervals or time units have equal length. Failure at time t or in interval $[a_{t-1}, a_t)$, respectively, is denoted by $T = t$ and the discrete hazard function of failure is given by the conditional probability

$$\lambda(t|x) \quad = \quad P(T = t | T \geq t, x)\,.$$

As demonstrated in Chapter 2, discrete survival data can be fit by binary regression models (binomial models) when considering for each individual its survival experience $y_n(t)$ through time, which is given by a sequence of binary outcomes with value zero when surviving t and value one when experiencing an event. The model can then be written in the general form

$$\lambda(t|x) = P\big(y(t) = 1|x\big) = h(\eta_t)\,.$$

The function $h : I\!\!R \to [0,1]$ is the response function defining the relation between the predictor η_t and the hazard $\lambda(t|x)$. The natural choice for h is the logistic response function (see Mc Cullagh & Nelder, 1983) yielding the logit model

$$\lambda(t|x) \quad = \quad \frac{\exp(\eta_t)}{1 + \exp(\eta_t)} \tag{4.6}$$

as given in (2.13), with the time-dependent predictor η_t. For ease of notation it is again assumed that interest is about a single covariate x with a dynamic effect and a single covariate $\tilde{x}$ with a constant effect. Then the predictor η_t has the semi-dynamic form

$$\eta_t = \gamma_t + \beta(t)x + \beta\tilde{x}\,, \tag{4.7}$$

with a baseline effect γ_t, a dynamic covariate effect $\beta(t)$ for covariate x, and a constant effect β for covariate $\tilde{x}$. Predictor (4.7) can be easily extended to a multivariate, semi-dynamic predictor by matrix representation, i.e.

$$\eta_t = \gamma_t + \mathbf{x}'\beta(t) + \tilde{\mathbf{x}}'\beta \tag{4.8}$$

where $\mathbf{x}$ comprises all covariates with a time-varying effect while the effects of the covariates in $\tilde{\mathbf{x}}$ remain constant over time. Note that the covariates in (4.8) and (4.7) may also be time-dependent.

Equivalently to (4.6) the observation model can be expressed by the likelihood given the model parameters. For the dynamic logit model it has the form

$$l = \prod_t \prod_{n \in R_t} \left[\frac{\exp\big(\gamma_t + \beta(t)x_n + \beta\tilde{x}_n\big)}{1 + \exp\big(\gamma_t + \beta(t)x_n + \beta\tilde{x}_n\big)} \right]^{y_n(t)} \cdot \left[\frac{1}{1 + \exp\big(\gamma_t + \beta(t)x_n + \beta\tilde{x}_n\big)} \right]^{(1-y_n(t))}$$

where as before R_t denotes the set of individuals at risk at time t. See Section 2.1.3 for further details.

The Parameter Model

In the semi-dynamic predictor (4.8) of the observation model, the baseline effect γ_t and the dynamic covariate effect $\beta(t)$ both vary with time. The development of these dynamic effects, or more precisely, the sequences γ_t and $\beta(t)$, $t = 1, 2, \ldots, K$, are defined in the parameter model. A common model for the transition of an effect $\beta(t)$, say, is given by a second order random walk (RW2)

$$\beta(t) = 2\beta(t-1) - \beta(t-2) + \varepsilon_t, \tag{4.9}$$

with i.i.d. Gaussian errors $\varepsilon_t \sim N(0, \sigma^2)$. The second order random walk can be seen as a penalisation of the roughness of the effect functions, where local deviations from the straight line are penalised. The variance parameter σ^2 controls the degree of smoothness, where a smaller variance will produce a smoother shape. For $\sigma^2 = 0$ the effects are restricted to be time-constant. Thus, from a Bayesian viewpoint, the parameter model represents the 'smoothness prior' of the dynamic model.

Equivalently to (4.9), the parameter model can be written in matrix notation by expressing the second order random walk in form of a general linear Gaussian transition model of the form

$$\boldsymbol{\beta} \sim N(0, \mathcal{K}_\beta^- \sigma^2),$$

where $\boldsymbol{\beta} = (\beta(1), \ldots, \beta(K))'$ and $\mathcal{K}_\beta^-$ is a generalised inverse of the *penalty matrix* $\mathcal{K}_\beta$, which for a second order random walk has the form

$$\mathcal{K}_\beta = \begin{pmatrix} 1 & -2 & 1 & & & & \\ -2 & 5 & -4 & 1 & & & 0 \\ 1 & -4 & 6 & -4 & 1 & & \\ & \ddots & \ddots & \ddots & \ddots & \ddots & \\ & & 1 & -4 & 6 & -4 & 1 \\ 0 & & & 1 & -4 & 5 & -2 \\ & & & & 1 & -2 & 1 \end{pmatrix} \tag{4.10}$$

Other common transition models are random walks of first order, generally yielding more coarse functions, the local linear trend model which includes an additional state process, or autoregressive models of higher order or with non-Gaussian errors. For details see Fahrmeir (1994). Alternatives to discrete transition models are for example smoothness prior specifications based on basis functions such as regression splines (Biller, 2000) or P-splines (Lang & Brezger, 2001a). However, in the context of *discrete* survival time, the representation (4.9) seems most suitable, as it defines the transition of the effects from one time point to the next by a random walk. It is used throughout this chapter for all dynamic covariate effects as well as for the baseline effects. When the discrete survival time is non-equidistant, random walk priors have to be modified appropriately to account for the non-equal length of the intervals $\tau_t = a_t - a_{t-1}$. For the second order random walk this can be achieved by specifying

$$\beta(t) = (1 + \tfrac{\tau_t}{\tau_{t-1}})\beta(t-1) - \tfrac{\tau_t}{\tau_{t-1}}\beta(t-2) + \varepsilon_t, \qquad (4.11)$$

with the Gaussian error $\varepsilon_t \sim N(0, \upsilon_t \sigma^2)$ and appropriate weights υ_t. A common weight is $\upsilon_t = \tau_t$, which is used in the following. More general weights have been proposed for example by Knorr-Held (1997).

Priors of Constant Coefficients

For a constant effect β of the semi-dynamic predictor (4.7) a Gaussian prior $\beta \sim N(0, c)$ is usually used. Here it is set $c \to \infty$, resulting in a non-informative (improper) prior which represents absolute prior ignorance about β.

The Hyperparameters

The dynamic effect model of state-space structure defined by the observation model and the parameter model contains a number of hyperparameters. The most important one is the variance σ^2 of the transition error. It regulates roughness of the dynamic effect function. Other hyperparameters involved in the second order random walk are the initial states γ_1, γ_2 and $\beta(1), \beta(2)$, respectively. Note that in the model definitions regarded above solely the parameter model conditions on hyperparameters while in general also the observation model could incorporate further hyperparameters.

If precise prior information is available fixed values can be assiged to the hyperparameters. However, in practice one usually prefers to be flexible and rather considers the hyperparameters to be stochastic. The full Bayesian approach then demands the specification of additional prior assumptions for these hyperparameters in a further level. Here a vague (diffuse) prior is assigned to the random walk variance σ^2, which is given by a highly dispersed inverse Gamma distribution $IG(a, b)$. Initial states of the Gaussian second order random walk are usually specified by Gaussian priors; $\gamma_1, \gamma_2 \sim N(0, cI)$ and $\beta(1), \beta(2) \sim N(0, cI)$. Here again non-informative priors are assigned which correspond to $c \to \infty$.

The Posterior Distribution of the Dynamic Logit Model

The observation model, the parameter model and the additional prior specifications together form the Bayesian dynamic logit model for discrete survival data. The attractive aspect of the representation of dynamic models in state-space structure is that it clearly exposes the different hierarchical levels. In fully Bayesian inference, the posterior distribution of the entire model parameters θ is of interest. By the principle of inverse probability (4.3) this is proportional to the joint distribution of the observations y, the covariates x and the parameters θ, which is itself given by the product of the likelihood times the parameter priors. (Here it shall be assumed, that the covariates are non-stochastic.) In the dynamic logit model, the full parameter vector θ comprises the entire sequences of the baseline effect $\gamma_3, \ldots, \gamma_K$ and of the dynamic covariate effects $\beta(3), \ldots, \beta(K)$, together with the initial states γ_1, γ_2 and $\beta(1), \beta(2)$, the RW2-variances $\sigma_\gamma^2, \sigma_\beta^2$ and, if present, the time-constant effects β. The joint distribution under consideration is thus

$$P(y, x, \gamma_1, \gamma_2, \gamma_3, \ldots, \gamma_K, \beta(1), \beta(2), \beta(3), \ldots, \beta(K), \beta, \sigma_\gamma^2, \sigma_\beta^2). \qquad (4.12)$$

It becomes clear, that one faces a high dimensional modelling problem. Yet, in order to define the joint distribution a number of independence assumptions have to be taken into account, which lead to helpful simplifications of the joint distribution (4.12). They can be summarised by the following four statements (Fahrmeir & Lang, 2001):

A1 The conditional probability of an event at time t given the covariates and the current states γ_t and $\beta(t)$, is independent of past states and the RW2-variances $\sigma_\gamma^2, \sigma_\beta^2$.

A2 The individual event probabilities are conditionally independent within a risk set R_t.

A3 The covariates and the censoring process are independent of past states.

A4 The current state of a second order random walk depends only on the two previous states while the initial states and the RW2-variance are mutually independent.

A more formal definition of these assumptions is given in Appendix A. To simplify notation, let α comprise the baseline effect as well as the dynamic covariate effect, i.e. $\alpha_t = (\gamma_t, \beta(t))'$, and $\boldsymbol{\alpha} = (\alpha_3', \ldots, \alpha_K')'$ and let Σ^2 be the diagonal matrix

$$\Sigma^2 = \begin{bmatrix} \sigma_\gamma^2 & \\ & \sigma_\beta^2 \end{bmatrix}.$$

By repeated application of the Bayes theorem and the independence assumptions **A1-A4**, the joint distribution of the dynamic logit model can be expressed as

$$
\begin{aligned}
P(y, x, \boldsymbol{\alpha}, \beta, \alpha_1, \alpha_2, \Sigma^2) \;=\; & \left\{ \prod_t \prod_{n \in R_t} l\big(y_n(t)|x_n, \alpha_t, \beta\big) \right\} \times \\
& \left\{ \prod_{t>2} \pi\big(\alpha_t | \alpha_{t-1}, \alpha_{t-2}, \Sigma^2\big) \right\} \times \\
& \pi(\beta) \times \pi(\alpha_1, \alpha_2) \times \pi(\Sigma^2),
\end{aligned}
$$

(Fahrmeir & Lang, 2001). Having chosen non-informative priors for the initial states and the constant effects, this reduces to

$$
\begin{aligned}
P(y, x, \boldsymbol{\alpha}, \beta, \alpha_1, \alpha_2, \Sigma^2) \;\propto\; & \left\{ \prod_t \prod_{n \in R_t} l\big(y_n(t)|x_n, \alpha_t, \beta\big) \right\} \times \\
& \left\{ \prod_{t>2} \pi\big(\alpha_t | \alpha_{t-1}, \alpha_{t-2}, \Sigma^2\big) \right\} \times \pi(\Sigma^2). \quad (4.13)
\end{aligned}
$$

For the semi-dynamic logit model with predictor (4.7) and Gaussian second order random walk priors, the joint distribution is explicitly given by

$$
\begin{aligned}
P(y, x, \boldsymbol{\alpha}, \beta, \alpha_1, \alpha_2, \Sigma^2) \;\approx\; & \prod_t \prod_{n \in R_t} \left[\frac{\exp(\eta_t)}{1+\exp(\eta_t)} \right]^{y_n(t)} \cdot \left[\frac{1}{1+\exp(\eta_t)} \right]^{(1-y_n(t))} \\
& \cdot \exp\left(-\frac{1}{2\sigma_\gamma^2} \gamma' \mathcal{K}_\gamma \gamma \right) \cdot \exp\left(-\frac{1}{2\sigma_\beta^2} \beta' \mathcal{K}_\beta \beta \right) \\
& \cdot \pi(\sigma_\gamma^2) \cdot \pi(\sigma_\beta^2). \quad\quad (4.14)
\end{aligned}
$$

Ignoring the priors $\pi(\sigma_\gamma^2)$ and $\pi(\sigma_\beta^2)$, the first three terms in (4.14) form a penalised likelihood with the penalisation defined by second order random walks. The penalisation terms have the familiar form which involves penalisation matrixes $\mathcal{K}_\gamma$ and $\mathcal{K}_\beta$

as defined in (4.10). The variance parameters σ_γ^2 and σ_β^2 act as smoothing parameters trading off between goodness of fit and roughness of the dynamic functions; the penalty increases or decreases as the variance becomes smaller or larger, respectively. (4.14) can be regarded as a discrete analogue to the penalised likelihood yielding continuous smoothing splines (Fahrmeir & Lang, 2001). Determining the posterior expectations of the variance parameters corresponds to a data driven estimation of the smoothing parameter.[1] Using Markov Chain Monte Carlo methods this can be realised jointly with the posterior estimation of the dynamic effects.

4.3 Markov Chain Monte Carlo Methods

Full Bayesian inference is based on the evaluation of the posterior distribution $P(\theta|y)$, or of posterior expectations of some function $f(\theta)$ as given in (4.4). The integrals involved in these expectations however, have caused serious limitations when applying Bayesian statistics in practice. In general they can not be calculated analytically. Note that the computation of the marginal probability of the data $P(y)$, which acts as a normalising constant of the posterior distribution, underlies the same difficulties. This problem particularly increases for complex models, such as the Bayesian dynamic logit model described above. To circumvent the difficult determination of these integrals Markov Chain Monte Carlo (MCMC) methods have been developed.

Markov Chain Monte Carlo is essentially Monte Carlo integration taking advantage of Markov Chain properties. Monte Carlo integration generally evaluates expectations or integrals of high dimensionality, such as (4.4), by efficiently drawing samples from the underlying distribution. In the Bayesian modelling framework MCMC is used to generate an 'empirical' posterior distribution of the parameters by drawing posterior samples $\theta^{(g)} \overset{a}{\sim} P(\theta|y)$, $g = 1, \ldots, G$. Markov chain theory ensures that under very general conditions the empirical distribution of the generated sample converges to the true posterior distribution. Thus, after a sufficiently long 'burn-in' phase during which the chain is levelling off, $\theta^{(1)}, \ldots, \theta^{(G)}$ can be considered as samples from the posterior distribution. Note that the sample is, however, dependent.

---

[1]Note that by definition of the parameter model, separate smoothing parameters are assigned to each of the dynamic effects, here γ and β.

The resulting empirical posterior distribution is used for inference on the data. Typically the empirical posterior mean of the model parameters θ is calculated as a point estimate, by

$$E_{\theta|y}[\theta] \approx \frac{1}{G} \sum_{g=1}^{G} \theta^{(g)} \, .$$

Also any other distribution characteristics of the parameters can be calculated from this sample, such as the standard deviation of the parameters or credibility regions. In addition the posterior distribution of any function of θ can be evaluated using this MCMC sample, by $f^{(g)}(\theta) = f(\theta^{(g)})$. E.g. the posterior expectation of a function $f(\theta)$ is approximated by the sample mean

$$E_{\theta|y}[f(\theta)] \approx \frac{1}{G} \sum_{g=1}^{G} f(\theta^{(g)}) \, .$$

MCMC sampling is based on full conditional distributions, which is the distribution of one component of θ given all remaining components:

$$P(\theta_s|\theta_1, \ldots, \theta_{s-1}, \theta_{s+1}, \ldots) := P(\theta_s|\theta_{-s}) \, .$$

New updates of the chain are iteratively generated from the full conditional distribution. Since the posterior distribution is uniquely determined by the set of its full conditional distributions the obtained sample is indeed a sample from the posterior distribution $P(\theta|y)$. Depending on the complexity of full conditional distribution different sampling schemes have been proposed.

The Gibbs sampling algorithm directly samples in each step from the full conditional distribution. However, the conditional distributions of the components themselves are often still quite complex and random numbers can not be directly drawn from them. The Hastings algorithm instead updates the chain by drawing from an arbitrary proposal density $P^{\star}(\theta_s \to \theta_s^{\star}|\theta_{-s})$ and accepts the generated random number $\theta_s^{\star}$ as a new update of the Markov chain with acceptance probability

$$p_a = \min \left\{ 1, \frac{P(\theta_s^{\star}|\theta_{-s}) \, P^{\star}(\theta_s^{\star} \to \theta_s|\theta_{-s})}{P(\theta_s|\theta_{-s}) \, P^{\star}(\theta_s \to \theta_s^{\star}|\theta_{-s})} \right\} \, . \tag{4.15}$$

Otherwise the chain remains in the current state θ_s. With (4.15) the acceptance probability depends on the current state of the chain θ_s, the proposed new state $\theta_s^{\star}$ and the current states of the other components θ_{-s}. Since only the ratio of the full conditional probabilities divided by the proposal probabilities is regarded in (4.15), implementation can be simplified by choosing a suitable proposal distribution. The

Metropolis-Hastings algorithm uses random walk proposals that are symmetric in θ_s and $\theta_s^\star$, i.e. $P^\star(\theta_s \to \theta_s^\star|\theta_{-s}) = P^\star(\theta_s^\star \to \theta_s|\theta_{-s})$.

The MCMC algorithm for the dynamic logit model

The attractive feature about Markov chain Monte Carlo methods is that they allow for hybrid algorithms. This means that the full conditional distribution of different components of the parameter vector θ may be updated by different algorithms. For dynamic survival modelling a Hastings step is used to update the states of the dynamic effects with random walk prior. It is based on the full conditional distribution

$$P(\alpha_t|y, x, \alpha_{-t}, \sigma^2) \propto \prod_{n \in R_t} l(y_n(t)|x_n, \alpha_t, \beta) \times P(\alpha_t|\alpha_{-t}, \sigma^2). \tag{4.16}$$

The second term of (4.16) is the conditional prior distribution of the states $P(\alpha_t|\alpha_{-t}, \sigma^2)$ given the neighbouring states and the current values of σ^2. This is used as a proposal distribution to draw new updates for the Markov Chain. The use of this 'conditional prior proposal' $P(\alpha_t|\alpha_{-t}, \sigma^2)$ yields an acceptance probability of the form

$$p_a = \min\left\{1, \frac{\prod_{n \in R_t} l(y_n(t)|x_n, \alpha_t^\star, \beta)}{\prod_{n \in R_t} l(y_n(t)|x_n, \alpha_t, \beta)}\right\}.$$

It can be viewed as the likelihood ratio at time t of the current sampling state to the new state. For the second order random walk given in (4.9) the conditional prior distribution is Gaussian and depends only on $\alpha_{t-2}, \alpha_{t-1}, \alpha_{t+1}, \alpha_{t+2}$ for $t > 2$, and on α_2, α_3 and $\alpha_1, \alpha_3, \alpha_4$ otherwise. Hence updating the states is straightforward. For further details see Knorr-Held (1999).

The variance parameter of the random walk prior σ_β^2 is updated in a Gibbs step. The full conditional distribution of σ_β^2 is $P(\sigma_\beta^2|\beta)$. It is again an inverse Gamma distribution $IG(\tilde{a}, \tilde{b})$ that depends solely on the current values of the dynamic states $\beta(1), \ldots, \beta(K)$. The parameters are obtained by $\tilde{a} = a + (k-2)/2$ and $\tilde{b} = b + \frac{1}{2}\sum \varepsilon_n^2$, where for equally spaced time-units ε is calculated as the observed error of the parameter model (4.9), i.e. $\varepsilon_t = \beta(t) - 2\beta(t-1) + \beta(t-2)$. For non-equally spaced time-units it is accordingly modified. The variance parameter of the baseline effect σ_γ^2 is updated the same way.

To generate a posterior sample of the constant effects β a Metropolis-Hastings step is employed, where the proposal distribution is chosen to be normal, i.e. $\beta^* \sim N(\beta, Q)$. Since a non-informative prior was assigned, the acceptance probability is again just based on the likelihood ratio, i.e.

$$p_a = \min \left\{ 1, \frac{\prod_{n \in R_t} l(y_n(t)|x_n, \alpha_t, \beta^*)}{\prod_{n \in R_t} l(y_n(t)|x_n, \alpha_t, \beta)} \right\}.$$

Methods to generate samples from standard distributions such as the normal distribution or the Inverse Gamma distribution are found in Ripley (1987).

In the algorithm described so far single components of the sequence of the state vector α are updated iteratively. To improve the performance of the algorithm with respect to convergence and to decrease computation time, instead $(\alpha_l, \ldots, \alpha_u)$ blocks can be updated simultaneously (Knorr-Held, 1999). The block-size should be neither too small nor too large and the block-boundaries should be randomly varied in each iteration in order to ensure better convergence at all states.

Mixing and tuning of the algorithm

Generally any inference obtained from an empirical posterior distribution is based on the assumption that the generated MCMC-sample sufficiently approximates the true posterior distribution. This in particular demands that the Hastings algorithm assures a good 'mixing', that is, that the Markov Chain properly covers the entire sample space. Mixing of a chain strongly depends on the overall acceptance rate. A very high overall acceptance rate indicates that the algorithm converges in relative small steps. On the other hand, if new replicates are rejected too often, the proposal distribution obviously offers inadequate samples and the chain does not change its state for a long period again causing slow convergence. In both cases the elements of the chain will show a high autocorrelation. Hence, one way to investigate the quality of a MCMC sample is to evaluate the autocorrelation function of the elements of the chain, where small values of the autocorrelation indicate good mixing. Other proposals for diagnosis of a MCMC-sample are found in Gilks et al. (1996) and particular in Gelman (1996). Evaluation of the acceptance rate and the autocorrelation function can be useful for tuning of the algorithm in preliminary runs. In the algorithm described for fitting a semi-dynamic logit model, tuning involves the starting values of the chain and the variance Q of the normal proposal distribution which is used to update the constant effects. Besides, as mentioned above, also block sizes of the dynamic effects influence convergence, as large blocks reduce the overall acceptance while smaller blocks lead to a higher acceptance rate. Practically for the algorithm described here the overall acceptance rate should lie between 0.3 and 0.7 (Fahrmeir & Knorr-Held, 1997). To further reduce the autocorrelation within the final sample used for posterior inference, the MCMC-output can additionally be thinned out by selecting only every r-th element. These strategies were pursued in this thesis.

Finally it should be noted that the posterior sampling by the proposed algorithm is computationally intensive, which is partially due to the particular structure of discrete survival data that was expressed as a binary time series. Another possible variation of the proposed model provides the probit model, where the response function used in the observation model has the form of a standard Gaussian distribution function. It allows to reduce the MCMC algorithm to a Gibbs sampler accelerating computation time.

4.4 Bayesian Model Diagnostic Tools

An essential task in statistical modelling is the verification of fitted models. This in particular implies the comparison of models of different complexity. Investigating and comparing the posterior means of model parameters, such as the covariate effects and their credibility regions, might sometimes not be satisfactory. Instead measures are required which explicitly allow validation of one model against an alternative. The classical proposal of Bayesian testing of two alternative models has been introduced by Jeffreys (1961) and relies on posterior model probabilities. However, this incorporates the marginal likelihood and therefore is accompanied by serious computational difficulties. In addition it is highly sensitive to prior specifications. Thus, alternatives seem worthwhile that circumvent these problems.

In this section, different criteria are studied that allow for model diagnostics and model comparison within the Bayesian framework. For most criteria the primary obstacle lies in their computation where additional difficulties result from the application of non-parametric (dynamic) modelling approaches and the special structure of survival data with censored observations. Therefore a major focus in the discussion of the criteria will be to study feasibility of their computation. Other important issues are practical aspects, such as implementation efforts and computation time. These are of particular interest within dynamic survival analysis, since high dimensional models with complex hierarchical structure are employed that involve a complex MCMC algorithm, as described above. It is thus essential that the model criteria themselves do not require too much additional computation effort. Altogether, criteria are preferred that can directly be determined from the empirical posterior distribution.

Naturally, the definition of model criteria is based on decision-theoretic and philosophical argumentation. Especially within the Bayesian context any proposed model criterion is therefore a controversial issue. The present section focuses on the formal definition of different criteria and a discussion of their practical applicability while a detailed insight into the background and the theoretical criticism will be given in Chapter 5.

4.4.1 The Posterior Model Probability

The posterior probability is generally considered to be the updated belief about a hypothesis, that is when the prior belief has been revised by the information gained from observed data. Hence, by following this basic concept of Bayesian statistics it seems plausible to evaluate a model's adequacy based on its model posterior probability (Jeffreys, 1961). Suppose the focus of interest is on the evaluation of model $\mathcal{M}$ for the data y with distribution $P(y)$. Assigning to each model a prior model probability $\pi(\mathcal{M})$, the posterior probability of model $\mathcal{M}$ given the observed data y is defined as

$$P(\mathcal{M}|y) = \frac{P(y|\mathcal{M}) \cdot \pi(\mathcal{M})}{P(y)}. \qquad (4.17)$$

The model prior $\pi(\mathcal{M})$ represents the subjective part in (4.17). It may express some additional penalisation assigned to a model. The probability $P(y)$ solely depends upon the data and acts as a normalisation constant. Hence, central focus in Bayesian posterior model diagnosis is put on $P(y|\mathcal{M})$ which is the marginal probability of the data given the model $\mathcal{M}$.

The Marginal Likelihood

To specify the marginal probability $P(y|\mathcal{M})$ suppose that under model $\mathcal{M}$ the data y have the likelihood $l(y|\theta, \mathcal{M})$, where θ is the vector of unknown model parameters for which the prior distribution $\pi(\theta|\mathcal{M})$ is specified. The marginal probability $P(y|\mathcal{M})$ can then be obtained by integrating the joint density $P(y,\theta|\mathcal{M})$ over the parameter space of θ, that is

$$P(y|\mathcal{M}) = \int_\Theta P(y,\theta|\mathcal{M})\,d\theta. \qquad (4.18)$$

Applying the Bayes theorem this can be transformed to

$$P(y|\mathcal{M}) = \int_\Theta l(y|\theta,\mathcal{M}) \cdot \pi(\theta|\mathcal{M})\,d\theta \qquad (4.19)$$

which is the conditional density of the observables $l(y|\theta,\mathcal{M})$ averaged with respect to the prior knowledge about the unobservables. $P(y|\mathcal{M})$ is also called the *marginal likelihood* or the *model likelihood*. From (4.19) it becomes clear that $P(y|\mathcal{M})$ is essentially the probability of observing the actual data y under model $\mathcal{M}$, calculated *before* any data became available and accordingly can be regarded as the *(prior) predictive probability of the data*. Jeffreys (1961) therefore states that "a formal Bayesian model adequacy criterion must evaluate the marginal density of the data at the actual observations."

4.4.2 The Bayes Factor

When models are compared in pairs the ratio of posterior model probabilities is considered, that is

$$\frac{P(\mathcal{M}_1|y)}{P(\mathcal{M}_0|y)} = \frac{P(y|\mathcal{M}_1)}{P(y|\mathcal{M}_0)} \cdot \frac{\pi(\mathcal{M}_1)}{\pi(\mathcal{M}_0)} \qquad (4.20)$$

The ratio of the model likelihoods

$$\mathrm{BF}_{1,0} = \frac{P(y|\mathcal{M}_1)}{P(y|\mathcal{M}_0)} \qquad (4.21)$$

defines the *Bayes factor* in favour of model $\mathcal{M}_1$ (and against model $\mathcal{M}_0$). A value $\mathrm{BF}_{1,0} > 1$ supports model $\mathcal{M}_1$, while a value $\mathrm{BF}_{1,0} < 1$ indicates, that model $\mathcal{M}_0$ is more plausible given data y. The odds of the model priors $\pi(\mathcal{M}_1)/\pi(\mathcal{M}_0)$ reflects the prior information on the model choice, depending on the prevailing situation and knowledge on the models. As mentioned above, it is the subjective part in model comparison. Consequently, the Bayes factor may be considered as representing 'objective' Bayesian model evaluation (Berger & Pericchi, 2001).

The methodology for quantifying the evidence of a model, respectively a 'scientific theory', by defining the Bayes factor was developed 1939 by Jeffreys in his book

'Theory of Probability', (Jeffreys, 1961, reprint of 1939). From (4.20) it emerges that the Bayes factor is the ratio of the posterior ratio to the prior odds

$$BF = \frac{\text{posterior ratio}}{\text{prior odds}}, \tag{4.22}$$

that is the factor with which the prior odds of models $\mathcal{M}_0$ and $\mathcal{M}_1$ is updated once data are observed. It can therefore be viewed as the weight of evidence for a model provided by the data. In the case of no prior preference, i.e. when all models are equally probable a priori with $\pi(\mathcal{M}_0) = \pi(\mathcal{M}_1) = 1/2$, the Bayes factor is equal to the posterior ratio.

When the model comparison is extended to a set of m models, say, the Bayes factor could again be used for model selection. Assuming equal priors Jeffreys suggests that the model which maximises the marginal likelihood $P(y|\mathcal{M})$ for the actual data y is to be selected. To rank a set of models it might be more practical to report for every model its 'normalised' model likelihood

$$\overline{P}(y|\mathcal{M}_j) = \frac{P(y|\mathcal{M}_j)}{\sum_{i=1}^{m} P(y|\mathcal{M}_i)}. \tag{4.23}$$

The Bayes factor of any pair of models from this set can then be calculated by BF $= \overline{P}(y|\mathcal{M}_j)/\overline{P}(y|\mathcal{M}_i)$. Alternatively one may report Bayes factors with respect to a reference model, e.g. the null model assigning a common mean to all observations.

The Bayes factor is the Bayesian counterpart to frequentists' (Fisherian) hypothesis testing and Jeffreys has even introduced his methodology as 'significance testing'. However, there are fundamental differences between these two concepts. First and foremost, they are of course based on opposing philosophical concepts of statistical inference. Moreover, in contrast to p-value testing, the Bayes factor offers a way to give evidence *in favour* of a model and symmetrically evaluates the null model and the alternative model. An additional important advantage is that the definition of the Bayes factor is very general and does not require alternative models to be nested, nor does it require any standard distributions. Moreover, beside model diagnosis and model comparison, the Bayes factor can also serve to calculate weights in Bayesian model averaging of a set of models. This is often used to account for model uncertainty (Draper, 1995).

Sometimes it is more convenient to consider $2\log(\text{BF})$, which is on the same familiar scale as the deviance and the likelihood ratio statistic. Jeffreys (1961) proposes an interpretation scheme of different values of the BF on the $\log_{10}$-scale, which Kass & Raftery (1995) transferred to the $2\log$-scale with slight modifications. Table 4.1

Table 4.1: Interpretation of the Bayes factor

$2\log(\text{BF}_{1,0})$	$\text{BF}_{1,0}$	Evidence	
< -10	< 0.007	'Very strong'	
-10 to -6	0.007 to 0.05	'Strong'	evidence for $\mathcal{M}_0$
-6 to -2	0.05 to $1/3$	'Positive'	
-2 to 2	$1/3$ to 3	'Not worth more than a bare mention'	
2 to 6	3 to 20	'Positive'	
6 to 10	20 to 150	'Strong'	evidence for $\mathcal{M}_1$
> 10	> 150	'Very strong'	

gives their categorisation, which is extended to the negative values emphasising the symmetry of the Bayes factor.

However, application of the Bayes factor is accompanied with serious technical problems and pitfalls. One reason the Bayes factor often has been criticised for is its dependency on prior specifications of the model parameters. It usually requires proper prior specifications for the model parameters to assure a sensible interpretation. Yet, in complex hierarchical modelling improper (non-informative) priors are often used. Even more problematic is the use of vague (diffuse) priors, as they result in the Lindley paradox. These problems are again discussed in more detail in Chapter 5. Beside the delicate dependency of the Bayes factor on prior specifications, a major practical difficulty is its computation, which causes a formidable problem in complex modelling. This issue is examined below.

Determination of the Bayes Factor

The major difficulty when applying the Bayes Factor is the determination of the marginal likelihood $P(y|\mathcal{M})$. Especially in high dimensional modelling problems, as e.g. in dynamic survival modelling, the marginal likelihood is neither explicitly available nor can it be evaluated analytically. Recalling the Bayes theorem (4.2) it becomes clear that the determination of $P(y|\mathcal{M})$ is the general obstacle in Bayesian modelling. It is the normalising constant of the posterior density of the parameters and thus essential for its explicit calculation. As illustrated in Section 4.3, in complex modelling situations Markov Chain Monte Carlo methods are usually employed to circumvent the difficult determination of $P(y|\mathcal{M})$ by generating an 'empirical' posterior distribution. MCMC methods are universal applicable to arbitrary complex models. This tempts to construct high dimensional hierarchical models which include different stochastic components at various levels. Besides dynamic effects, random effects or spatial effects for example, are additionally included. However, the validation of these models using the Bayes factor brings back the problem of determining the high dimensional integral defining the marginal likelihood $P(y|\mathcal{M})$. Hence, the universal applicability of MCMC leads to a 'Catch-22'.

Various methods have been proposed to eliminate the above problem. One approach is to derive estimates of the marginal likelihood based on approximations or numerical integration methods. Other approaches propose to derive the marginal likelihood from an empirical posterior model distribution generated by MCMC sampling. In the following an overview over the different proposals is given. A major focus is laid on the applicability of the methods for comparing dynamic survival of the form given in Section 4.2.

The Schwarz criterion (Bayesian Information Criterion)

The most common approximation to the Bayes factor is the Schwarz criterion (Schwarz, 1978). It is defined by

$$SC(\mathcal{M}_1, \mathcal{M}_0) = \log l(y|\hat{\theta}_1, \mathcal{M}_1) - \log l(y|\hat{\theta}_0, \mathcal{M}_0) + (df_0 - df_1)\log(N^*),$$

where $\hat{\theta}_j$ is the maximum likelihood estimate of the parameter vector θ_j under model $\mathcal{M}_j$, df_j is the dimension of θ_j and N^* is the number of independent observations. The Schwarz criterion gives a rough approximation to the logarithm of the Bayes Factor, i.e.

$$\frac{SC - \log(BF_{10})}{\log(BF_{10})} \xrightarrow{N^* \to \infty} 0,$$

Similarly, one often looks at minus twice the Schwarz criterion of model $\mathcal{M}$, which is known as the *Bayesian information criterion* BIC, that is

$$BIC(\mathcal{M}) = -2\log l(y|\hat{\theta}, \mathcal{M}) + df \log(N^*).$$

The BIC is on the same common scale as the deviance and has the structure of an Information criterion, where the goodness of fit measured by the deviance $-2\log l(y|\hat{\theta}, \mathcal{M})$ is penalised by $\log(N^*)$-times the model complexity. Hence, smaller values of the BIC indicate a better fit of the model, while for a larger sample size the penalisation of model complexity increases.

An advantage of the Schwarz criterion and the BIC might be that they do not require specification of prior distributions since they are based on maximum likelihood estimates. The major benefit is without doubt that they are easy to compute when maximum likelihood estimates are available. However, the calculation of the Schwarz criterion and the BIC requires the specification of the dimension df of θ, i.e. the effective number of parameters. This is no difficulty in parametric modelling, where the dimension of the model corresponds to the dimension of the unknown (independent) parameter vector. Unfortunately for non-parametric models the true model dimension is not clearly defined, so that these criteria can not be directly calculated. Moreover, when modelling survival data with censored observations, the actual 'sample size' N^* is unclear. Another problem is related to the goodness of the approximation, which is only satisfactory as long as the number of parameters involved in the comparison is reasonable small relative to the sample size (Gelfand & Ghosh, 2001). In dynamic modelling the number of parameters is, however, rather large.

The Laplace method

A common method to approximate integrals is the Laplace integration. It is based on a normal approximation of the argument of the integral. To approximate the integral of the marginal likelihood $\int l(y|\theta, \mathcal{M})\pi(\theta|\mathcal{M})$, it is assumed that the posterior density and the likelihood function are highly peaked. Taylor expansion of the logarithm of the joint distribution $\log l(y|\theta, \mathcal{M})\pi(\theta|\mathcal{M})$ about the posterior mode $\tilde{\theta}$ yields an approximation of the joint distribution which has the form of a normal distribution with mean $\tilde{\theta}$ and covariance matrix $\tilde{\Sigma} = [-H_{\tilde{\theta}}^2]^{-1}$, where $H_{\tilde{\theta}}^2$ is the Hessian matrix of the second derivatives of $\log l(y|\theta, \mathcal{M})\pi(\theta|\mathcal{M})$. Integration then yields in the approximation

$$P(y|\mathcal{M}) \approx (2\pi)^{df/2}|\tilde{\Sigma}|^{1/2}\, l(y|\tilde{\theta}, \mathcal{M})\pi(\tilde{\theta}|\mathcal{M})\,, \qquad (4.24)$$

where df is again the model dimension. In most situations computation of (4.24) should be feasible. Otherwise, a variant of the Laplace integral can be used, which is based on the maximum likelihood estimate $\hat{\theta}$ and the observed Fisher information (Kass & Raftery, 1995). For 'well behaved' cases in which the likelihood is not grossly non-normal, the approximation of the integral in (4.18) by (4.24) or its variants are sufficiently accurate to serve for model comparisons (Kass & Raftery, 1995). However, Laplace integration suffers from the same drawbacks as the BIC in that the goodness of the approximation depends on the ratio of the number of parameters to the sample size. And just as for the Schwarz approximation, Laplace approximations only allow for an unproblematic computation of the marginal likelihood if the model dimension is specified. This renders it inapplicable in non-parametric dynamic modelling.

Importance sampling

A different approach to estimate the marginal likelihood is via sampling based methods. The marginal likelihood $P(y|\mathcal{M}) = \int l(y|\theta, \mathcal{M})\pi(\theta|\mathcal{M})d\theta$ can be seen as the 'prior mean' of the likelihood. Hence it seems natural to estimate this mean using Monte Carlo integration methods. One general approach is to use importance sampling, where a sample $\theta^{(1)}, \theta^{(2)}, \ldots, \theta^{(G)}$ is drawn from some importance distribution $P^\star(\theta)$, (Kass & Raftery, 1995). Since the likelihood is a simple function of θ, this sample can be used to generate a sample of the likelihood $l(y|\theta^{(g)}, \mathcal{M})$, $g = 1, \ldots, G$, and in turn estimate the prior expectation of the likelihood by the weighted sample mean

$$P(y|\mathcal{M}) \approx \frac{\sum\limits_{g=1}^{G} w_g \cdot l(y|\theta^{(g)}, \mathcal{M})}{\sum\limits_{g=1}^{G} w_g}, \qquad (4.25)$$

with the importance weights $w_g = \pi(\theta^{(g)}|\mathcal{M})/P^\star(\theta^{(g)})$, (see e.g. Ripley, 1987). The importance distribution is any distribution which is not too far from $\pi(\theta|\mathcal{M})$ and allows for sampling of the parameter θ. One possible choice is to sample directly from the prior density $P^\star(\theta) = \pi(\theta|\mathcal{M})$, so that the weights equal one. However, for a high dimensional parameter space direct prior sampling might be impossible. Alternatively, if posterior analysis is based on MCMC methods it might be opportune to use the generated posterior sample instead. Substituting $P^\star(\theta) = P(\theta|y, \mathcal{M})$ in equation (4.25), yields in

$$P(y|\mathcal{M}) \approx \Big[\frac{1}{G} \sum_{g=1}^{G} l(y|\theta^{(g)}, \mathcal{M})^{-1}\Big]^{-1} \qquad (4.26)$$

which is the harmonic mean of the likelihood values over the posterior distribution. The harmonic mean (4.26) converges almost surely to the correct value of $P(y|\mathcal{M})$ as G increases (Newton & Raftery, 1994). However this approximation is unstable. This is caused by the occasional occurrence of sample values $\theta^{(g)}$ with a small likelihood value which have a large effect on the harmonic mean. Nonetheless, the accuracy of its approximation is presumed to be sufficient for model comparison based on interpretation schemes, as that of Table 4.1, (Kass & Raftery, 1995).

Direct estimation of the model likelihood via MCMC sampling

Chib (1995) presents an approach to directly estimate the marginal likelihood of a model using the Bayes theorem. Noting that by the Bayes theorem (4.2) the marginal likelihood is given by

$$P(y|\mathcal{M}) = \frac{l(y|\theta, \mathcal{M}) \cdot \pi(\theta|\mathcal{M})}{P(\theta|y, \mathcal{M})}, \quad (4.27)$$

he proposes to estimate the marginal likelihood by evaluating (4.27) at some high density point $\breve{\theta}$. Assuming that the likelihood $l(y|\breve{\theta}, \mathcal{M})$ and the prior $\pi(\breve{\theta}|\mathcal{M})$ can directly be calculated, determination of the marginal likelihood is reduced to finding an estimate of the posterior ordinate $P(\breve{\theta}|y, \mathcal{M})$. As a high density point $\breve{\theta}$ we might, for example, choose the posterior mean resulting from MCMC sampling.

Chib's approach employs block-wise sampling where the parameter space is divided into B blocks. The algorithm then generates $B-1$ Markov chains by running the Gibbs sampler. In each run, an additional block of parameters is fixed at the high density values $\breve{\theta}$ obtained by the previous run, while the remaining parameters are sampled conditionally on these fixed blocks.

Suppose the parameter vector θ is divided into B blocks, i.e. $\theta = (\theta_1, \ldots, \theta_B)$. The posterior ordinate of $\breve{\theta}$ can be split up into

$$P(\breve{\theta}|y) = P(\breve{\theta}_1|y) P(\breve{\theta}_2|\breve{\theta}_2, y) \cdots P(\breve{\theta}_B|\breve{\theta}_1, \ldots, \breve{\theta}_{B-1}, y). \quad (4.28)$$

In the Gibbs sampling approach each factor of (4.28) can directly be determined by averaging the full conditional densities

$$P(\breve{\theta}_s|\breve{\theta}_1, \ldots, \breve{\theta}_{s-1}, y) = \frac{1}{G} \sum_{g=1}^{G} P(\breve{\theta}_s|\breve{\theta}_1, \ldots, \breve{\theta}_{s-1}, \theta_{s+1}^{(g)}, \ldots, \theta_B^{(g)}, y),$$

for $s = 2, \ldots, B-1$, where $\theta_{s+1}^{(g)}, \ldots, \theta_B^{(g)}$ are obtained from reduced Gibbs sampling run in which $\theta_1, \ldots, \theta_{s-1}$ are fixed.

However, as mentioned above, in complex modelling problems, like e.g. in dynamic regression modelling, direct Gibbs sampling from full conditional distributions is usually not feasible. Chib & Jeliazkov (2001) have recently extended the approach to Metropolis-Hastings sampling yielding a more general applicability. They have presented an algorithm to determine the posterior ordinate, which efficiently uses the results for previous runs. Let $P^\star(\theta_s \to \theta_s^\star | \theta_{-s})$ denote the proposal density of a Metropolis-Hastings algorithm and p_a be the acceptance probability for moving from the actual state θ_s to the proposed state $\theta_s^\star$. The reduced conditional ordinates can then be determined by

$$P(\breve{\theta}_s | \breve{\theta}_1, \ldots, \breve{\theta}_{s-1}, y) = \frac{E_1[p_a(\theta_s \to \breve{\theta}_s | \theta_{-s}) \, P^\star(\theta_s \to \breve{\theta}_s | \theta_{-s})]}{E_2[p_a(\breve{\theta}_s \to \theta_s | \theta_{-s})]},$$

where E_1 is the expectation with respect to the reduced conditional distribution $P(\theta_s, \ldots, \theta_B | y, \breve{\theta}_1, \ldots, \breve{\theta}_{s-1},)$, and E_2 is the expectation with respect to $P(\theta_{s+1}, \ldots, \theta_B | y, \breve{\theta}_1, \ldots, \breve{\theta}_s) P^\star(\breve{\theta}_s \to \theta_s | \theta_{-s})]$, where $\theta_{-s} = \breve{\theta}_1, \ldots, \breve{\theta}_{s-1}, \theta_{s+1}, \ldots, \theta_B$. These expectations can be calculated from the output of reduced Metropolis-Hastings runs by

$$E_1 = \frac{1}{G} \sum_g p_a(\theta_s^{(g)} \to \breve{\theta}_s | \theta_{-s}^{(g)}) \, P^\star(\theta_s^{(g)} \to \breve{\theta}_s | \theta_{-s}^{(g)})$$

where $\theta_{-s}^{(g)} = \breve{\theta}_1, \ldots, \breve{\theta}_{s-1}, \theta_{s+1}^{(g)}, \ldots, \theta_B^{(g)}$ and $\theta_s^{(g)}, \ldots, \theta_B^{(g)}$ result from a reduced Metropolis-Hastings run with $\theta_1, \ldots, \theta_{s-1}$ fixed. Analogously,

$$E_2 = \frac{1}{J} \sum_j p_a(\breve{\theta}_s \to \theta_s^{(j)} | \theta_{-s}^{(j)})$$

where $\theta_{s+1}^{(g)}, \ldots, \theta_B^{(g)}$ result from the reduced run with $\theta_1, \ldots, \theta_s$ fixed. Note that this run is equivalent to the next reduced run where ordinate $P(\breve{\theta}_{s+1} | \breve{\theta}_1, \ldots, \breve{\theta}_s, y)$ is calculated. Additionally, $\theta_s^{(j)}$ is drawn from the proposal density $P^\star(\breve{\theta}_s \to \theta_s | \theta_{-s}^{(j)})]$. Once the posterior ordinates are calculated, the estimation of the marginal likelihood can be directly performed using the Bayes theorem.

A major advantage of these two proposals is that they are based on the same full conditional distributions that are used for posterior parameter estimation and therefore only require very little additional implementation and tuning effort. However, since $B - 1$ MCMC-runs are required the algorithms are computationaly very expensive.

MCMC methods to determine the posterior model probability

Recently, approaches became popular which employ MCMC methods to directly generate the posterior model probability for an entire set of models $\mathcal{M}_1, \ldots, \mathcal{M}_m$ by jointly sampling over the model space $\mathcal{M}$ and parameter space Θ. These approaches deliver posterior model probabilities according to the frequencies with which the algorithm 'visits' each model. Assuming equal model priors these frequencies directly provide the normalised Bayes factors of the form (4.23). Otherwise the Bayes factor of each model can be calculated as the ratio of its posterior probability and its model prior probability.

The sampling algorithm suggested in Carlin & Chib (1995) operates over the full product space $\mathcal{M} \times \prod_{j \in \mathcal{M}} \Theta_j$. In every step it draws a model $\mathcal{M}_j$, $j \in \{1, \ldots, m\}$ together with the entire set of parameters of all models $\theta_1, \ldots, \theta_m$. Since this requires complete definition of the joint model specification they propose the use of pseudo-priors for parameters θ_i of the models $\mathcal{M}_i \neq \mathcal{M}_j$. These pseudo-priors can be arbitrary but should be chosen in a way that ensures a proper mixing of the algorithm.

An alternative strategy is given by a reversible jump algorithm (Green, 1995), which operates in the union space $\mathcal{M} \times \cup_{j \in \mathcal{M}} \Theta_j$. This algorithm allows to jump between models with parameter spaces of different dimension. Instead of sampling in each step from the entire parameter space $\prod_{j \in \mathcal{M}} \Theta_j$ it only samples from the actual parameter space Θ_j which avoids pseudo-prior sampling and is therefore faster.

These model posterior sampling methods are attractive, as they are as generally applicable as MCMC sampling methods are. However, theoretically these approaches assume that the set of models under consideration includes the true model. Practically this means that these algorithms validate each model conditional on $\mathcal{M}$ and therefore only allow for comparison within this prespecified model space $\mathcal{M}$. Hence, these sampling methods may be of advantage if the set of models under consideration is prefixed or arises naturally from the modelling context and is rather large. For example they are used in Bayesian regression smoothing where different models are fit with varying number of knots (Biller, 2000). In these situations the resulting Bayes factor may even serve as a weight for model averaging. Due to conditioning on the prespecified model space, these algorithms do not yield estimates for the marginal probability but relative measures, i.e. Bayes factors of the form (4.23) and hence they do not allow for (independent) diagnosis of a single model. In particular, this implies that for the comparison of two models both must be re-fit in a joint algorithm. In practice this is a severe drawback, if, for example, two scientists independently fit different models and wish to compare them retrospectively.

Note that moreover adequate posterior samples $\theta^{(g)}$, $g = 1, \ldots, G$ are only produced for those models which have been visited sufficiently often during the algorithm.

Hence, the sample generated for models with a small posterior model probability may not comprise much information on the parameter posterior and might therefore be unusable for parameter estimation. This problem might have particular impact when using these algorithms for (automatic) model averaging.

Applicability of the methods to dynamic survival modelling

The general difficulties of the outlined methods to compute Bayes factors have already been mentioned. When applying these methods to our situations additional obstacles arise from the special context of non-parametric dynamic survival modelling. One substantial peculiarity is that in non-parametric modelling the true model dimension, i.e. the effective number of parameters, is unclear. In addition, similar to random effect models, in dynamic survival models the ratio between the number of parameters and the number of observations is usually rather high, which affects adversely the goodness of the approximations. Last but not least, when arguing for a method, practical aspects should also be considered that influence their applicability to compare dynamic survival models which includes in particular implementation and computation. When survival data are viewed as sequences of binary outcomes and fitted by time-series regression models as the dynamic logit model of Section 4.2, posterior parameter estimation is already rather time consuming and any additional increase of computation time must be avoided whenever possible. Besides, also the implementation effort of the methods should be taken into account, to ensure its application in practice. In the following the different proposals are compared with respect to these requirements to investigate which of them are practically applicable in dynamic survival analysis.

When the number of effective model parameters is unknown, as in no-parametric modelling, direct approximations of the Bayes factor by the BIC or Laplace Integration can not be determined. Instead sampling based methods have to be used to compute the marginal likelihood. However, these methods are known to be less precise (Kass & Raftery, 1995) and computationally far more demanding.

The sampling-based proposal of Chib & Jeliazkov (2001) employs a block-wise Metropolis-Hastings sampler drawing from the same full conditional distributions that are used for posterior parameter estimation. Since posterior estimation of the dynamic logit model is already accomplished by block-wise sampling, as outlined in Section 4.3, implementation can be based on adapting the existing computer codes, by allowing to partially fix the parameters. Here the independence assumptions of Section 4.2 are especially helpful. In particular no extra tuning or preliminary runs are needed. These facts save a considerable amount of time and effort. A major drawback is that the algorithm of Chib & Jeliazkov clearly requires substantial computation time, which is a serious burden in the context of dynamic survival modelling. Moreover, as discussed in Section 4.3, when using a block-wise Metropolis-Hastings algorithm splitting up γ and β the block-sizes can not be chose arbitrarily, as they influence the acceptance rate of the algorithm and thus the convergence of the chain. In fact, often rather small blocks are required to ensure a good mixing. This in turn increases the total number of blocks and consequently the number of Metropolis-Hastings runs needed in the algorithm of Chib & Jeliazkov. Apart from the choice of block-sizes, it might also be problematic that the block-wise algorithm for dynamic effect models shows bad mixing at the block-bounds. For this reason it is proposed to vary the block-bounds in each iteration step (compare Section 4.3). In the algorithm of Chib & Jeliazkov the bounds are fixed, however. In general further investigations are needed, which examine the performance of this algorithm and in particular study the impact of badly mixing chains.

The algorithm of Carlin & Chib and the reversible jump algorithm generate posterior model probabilities by jointly sampling over the model space and the parameter space. This only seems to be practical when the complete set of models of interest can be prespecified in advance and any later extension of the model class is excluded. Moreover, as mentioned above, for models with a small posterior model probability the generated chain might show rather bad convergence and might not allow appropriate posterior inference. In addition, these algorithms demand considerable implementation effort and careful tuning to assure acceptable approximation of the posterior distribution. This especially applies for high dimensional models where preliminary runs may be needed.

Han & Carlin (2000) give a comparative review over a number of MCMC sampling based methods. This includes the algorithm of Chib and Chib & Jeliazkov for marginal likelihood estimation as well as the product-space sampler of Carlin & Chib and different versions of the reversible jump algorithm. They study the performance of these algorithms in two applications where they particularly inspect practical handling, computation time and convergence behaviour. They detect convergence problems of those algorithms jointly sampling over the model and parameter space, when applied to complex modelling situations. The Gibbs-sampler based algorithm for marginal likelihood estimation of Chib shows a better behaviour in their applications.

Finally it can be concluded that if MCMC methods are used for posterior estimation of the model parameters, the most appealing approach to obtain the marginal likelihood is given by importance sampling from the posterior distribution which results in the harmonic mean. Its determination can be based on the generated posterior sample which is directly available. All other sampling based methods require formidable implementation, tuning and computation effort and are therefore not practical in application. Or, to put it in the words of Han & Carlin (2000): "Less formal Bayesian model choice methods may offer a more realistic alternative in many cases." In the following I will investigate some alternatives.

4.4.3   The Posterior Bayes Factor

As mentioned above, the Bayes factor is very sensitive to prior specifications. Generally, the sensitivity of model diagnosis to the specified priors may not be a matter of concern if these priors accurately represent one's subjective belief about the parameters. In many situations, however, one would rather prefer a statistical conclusion which is insensitive to prior specifications. Especially when the available prior information is vague and highly dispersed priors have been chosen to represent prior ignorance, the use of the Bayes factor may be problematic. In particular, vague priors are cause of the Lindley paradox. Yet, in applied Bayesian statistics the use of vague priors is well established. This issue is discussed in more detail in Chapter 5.

The sensitivity of the Bayes factor (4.21) to the prior specification results from the construction of the marginal probability $P(y|\mathcal{M})$ as a *prior* mean of the likelihood, that is the likelihood $l(y|\theta, \mathcal{M})$ averaged against the prior $\pi(\theta|\mathcal{M})$. To form a measure of model diagnosis that is robust toward prior specifications Aitkin (1991) proposes instead to evaluate the *posterior* mean of the likelihood. This yields the definition of the *posterior Bayes factor* for comparison of model $\mathcal{M}_0$ and $\mathcal{M}_1$:

$$\text{PBF}_{1,0} = \frac{\bar{l}(y|\mathcal{M}_1)}{\bar{l}(y|\mathcal{M}_0)} \tag{4.29}$$

where the ratio is taken of the posterior expectations of the likelihood, i.e.

$$\bar{l}(y|\mathcal{M}) = \int_{\Theta|y} l(y|\theta, \mathcal{M}) \cdot P(\theta|y, \mathcal{M})d\theta = E_{\theta|y}\left[l(y|\theta, \mathcal{M})\right]. \tag{4.30}$$

For given data y the likelihood $l(y|\theta, \mathcal{M})$ is a function of the parameters θ. Thus, the posterior distribution of the likelihood is directly derived from the posterior distribution of θ. In particular, if posterior sampling methods such as MCMC are used, a posterior sample of $l(y|\theta,\mathcal{M})$ can directly be constructed from the generated posterior sample $\theta^{(g)}$, and the posterior expectation of the likelihood is obtained by the sampling mean

$$\bar{l}(y|\mathcal{M}) = \frac{1}{G}\sum_{g=1}^{G} l(y|\theta^{(g)},\mathcal{M}).$$

With the Bayes theorem (4.2) the posterior distribution of θ is given by $P(\theta|y,\mathcal{M}) = l(y|\theta,\mathcal{M})\pi(\theta,\mathcal{M})/\int l(y|\theta,\mathcal{M})\pi(\theta,\mathcal{M})d\theta$ and the posterior mean of the likelihood (4.30) can equivalently be written as

$$\bar{l}(y|\mathcal{M}) = \frac{\int l(y|\theta,\mathcal{M})^2 \cdot \pi(\theta|\mathcal{M})d\theta}{\int l(y|\theta,\mathcal{M}) \cdot \pi(\theta|\mathcal{M})d\theta}.$$

This can also be viewed as a weighted average where the normalised likelihood $l(y|\theta,\mathcal{M})/\int l(y|\theta,\mathcal{M})\pi(\theta|\mathcal{M})d\theta$ serves as weight.

Aitkin gives a justification for the posterior Bayes factor based on prediction theoretic considerations. When y_{rep} denote replicated data that comes from the same model as the actual data y, the (posterior) predictive distribution of observing y_{rep} has the form

$$P(y_{rep}|\theta,y,\mathcal{M}) = \int l(y_{rep}|\theta,\mathcal{M}) \cdot P(\theta|y,\mathcal{M})d\theta.$$

This is equivalent to the posterior mean of the likelihood when substituting y_{rep} by the observed data y. Thus, the posterior mean of the likelihood can be interpreted as the predictive probability of new data, which has the same value of the sufficient statistics as the observed data, i.e. for which $l(y_{rep}|\theta,\mathcal{M}) = l(y|\theta,\mathcal{M})$.

Since the posterior Bayes factor is based on the posterior mean, it is less sensitive to prior specifications and even well defined if improper priors are used. In particular the posterior Bayes factor is not subject to the Lindley paradox. With definition (4.29), the posterior Bayes factor is symmetric and does not require the models to be nested as the criterion is constructed from the two likelihood expectations $\bar{l}(y|\mathcal{M}_0)$ and $\bar{l}(y|\mathcal{M}_1)$ which result from separate posterior distributions. Large values of $\text{PBF}_{1,0}v$ indicate evidence in favour for model $\mathcal{M}_1$.

Aitkin (1991) proposes an interpretation scheme for the posterior Bayes factor similar to Jeffreys interpretation of the Bayes factor. It is summarised in Table 4.2. Note that the values of the posterior Bayes factor are, however, generally not on the same level as the Bayes factor. In a proceeding article (Aitkin, 1997), Aitkin shows that the posterior Bayes factor requires substantial re-calibration to provide convincing

Table 4.2: Interpretation of the Posterior Bayes factor

$2\log(\text{PBF}_{1,0})$		$\text{PBF}_{1,0}$		Evidence	
6	to 9	20	to 100	'Strong'	
9	to 14	100	to 1000	'Very Strong'	for $\mathcal{M}_1$
	> 14		> 1000	'overwhelming'	

evidence against a hypothesis. Instead of regarding the ratio of two posterior means, he therefore studies the posterior distribution of the likelihood ratio LR and consequently restricts his considerations to nested hypothesis problems. This issue will be reviewed in more detail in Chapter 5.

In a discussion of the posterior Bayes factor Dempster (1997b) questions the consideration of the posterior mean of the *likelihood* for model comparison and mentions that "it might have seemed more natural to work with the logarithm of the LR and use its mean as a summary". This leads to the Bayesian deviance, which is studied in the next section.

4.4.4 The Model Deviance

As we have seen in Chapter 3, within the framework of classical likelihood-based statistics, model comparison of two nested models $\mathcal{M}_0$, $\mathcal{M}_1$ with $\Theta_0 \subset \Theta_1$ is traditionally performed using the generalised likelihood ratio test

$$LR_{\hat{\theta}}(y) = \frac{l(y|\hat{\theta}_0)}{l(y|\hat{\theta}_1)}, \qquad (4.31)$$

where $\hat{\theta}_j$ is the maximum likelihood estimate of the particular model. It is known, that for suitable regularity conditions such as asymptotic normality and the parameter space Θ_0 lying properly in Θ_1, the likelihood ratio statistic $-2\log LR_{\hat{\theta}}(y)$ has an asymptotic χ^2_{df}-distribution, where df is the difference in dimensionality of the parameter spaces, that is $df = \dim(\Theta_1) - \dim(\Theta_0)$. Small values of $LR_{\hat{\theta}}(y)$ suggest rejection of model $\mathcal{M}_0 : \theta \in \Theta_0$ in favour of model $\mathcal{M}_1$. From (4.31) it results that a general measure for the goodness of fit of a model can be given by the likelihood ratio statistic between the model under consideration and the saturated model, explaining all the variation of the data. This yields in the definition of the classical maximum likelihood deviance

$$D(\hat{\theta}|\mathcal{M}) = -2\log l(y|\hat{\theta}, \mathcal{M}) + 2\log l(y), \qquad (4.32)$$

(McCullagh & Nelder, 1983). By definition the saturated model maximises the likelihood, i.e. $l(y)$ is the global maximum of the likelihood. Since it is a function of the data only, it acts as a standardising term (saturated standardisation). As it has no impact on model selection it is often omitted. Intuitively, large values of $D(\hat{\theta}|\mathcal{M})$ indicate less support of model $\mathcal{M}$. Note that the deviance (4.32) is a relative measure for model comparison and not suitable to formally test the adequacy of a model.

In the other extreme case there is the null model, that is the model consigning a common mean to all observations. This model also has an important role in the context of model rating. The gain in the goodness of fit of a model is often reported by the reduction in deviance of the fitted model with respect to the null model (null standardisation). If $\mathcal{M}_0$ denotes the null model this yields

$$D(\hat{\theta}_0|\mathcal{M}_0) - D(\hat{\theta}|\mathcal{M}) = -2\log l(y|\hat{\theta}_0, \mathcal{M}_0) + 2\log l(y|\hat{\theta}, \mathcal{M}),$$

where large values point to a large improvement in the goodness of fit by model $\mathcal{M}$. It is useful to note that assuming a standard setting, the saturated model has dimension N^*, where N^* is the number of independent observations in y. On the other hand, the null model is of dimension one. Thus, any model $\mathcal{M}$ lying between these two extremes has dimensionality $1 \leq df_{\mathcal{M}} \leq N^*$.

While the classical maximum likelihood deviance is based on the log-likelihood evaluated at the maximum likelihood estimates $\hat{\theta}$, in the Bayesian perspective the parameters θ are random variables. This implies that the Bayesian definition of the deviance is a function of θ, which is

$$D(\theta|\mathcal{M}) = -2\log l(y|\theta, \mathcal{M}) + 2\log l(y).$$

As above, $\log l(y)$ is a standardising term which has no impact on model choice and is therefore often neglected. Since for given data y the deviance $D(\theta|\mathcal{M})$ is a function of the parameters θ, its posterior distribution can be derived directly from the posterior distribution of the parameters $P(\theta|y, \mathcal{M})$. In particular, if MCMC methods are applied, the posterior sample of the parameters $\theta^{(g)}$ can be used to form a posterior sample of the deviance $D^{(g)} = D(\theta^{(g)}|\mathcal{M})$.

Dempster (1997b) suggests to perform model comparison by regarding the posterior distribution of the deviance, or by plotting the posterior density of the log-likelihood under each competing model. These can be easily obtained by kernel density estimation using the output of MCMC sampling. However, when several models fit the data equally well these kernel density estimates might be hard to distinguish. In this case one-point measures summarising goodness of fit might be more convenient. Dempster (1997a) therefore proposes the use of the posterior mean of the log-likelihood for model comparison, or equivalently, the posterior expectation of the deviance, that is

$$E_{\theta|y}[\theta] = \int -2\log l(y|\theta, \mathcal{M}) \cdot P(y|\theta, \mathcal{M}) d\theta.$$

It can be derived from the MCMC sample $\theta^{(g)}$ as the sample mean

$$E_{\theta|y}[D(\theta|\mathcal{M})] \approx \overline{D}(\mathcal{M}) = \frac{1}{G} \sum_{g=1}^{G} D(\theta^{(g)}|\mathcal{M}).$$

Also Spiegelhalter et al. (2001) employ the posterior mean of the deviance as a measure of goodness of fit for model comparison. For an informal justification they consider the prior predictive expectation of the posterior expected deviance, i.e. $E_y[\overline{D}]$. This represents the posterior mean of the deviance expected *before* any data became available.

Alternatively, in accordance with (4.32), the goodness of fit of a model $\mathcal{M}$ may be summarised by the deviance of the posterior expectation of the model parameters, that is

$$D(E_{\theta|y}[\theta]|\mathcal{M}) = D(\bar{\theta}|\mathcal{M}).$$

If $\bar{\theta}$ is used for model construction this can be interpreted as the goodness of fit of the 'final' model. Spiegelhalter et al. (2001) call $D(\bar{\theta}|\mathcal{M})$ a 'plug-in estimate of fit'.

In the dynamic logit model for survival data the deviance is derived from the representation (2.14) (Chapter 2) of its likelihood, regarding for each individual n its survival experience $y_n = (y_n(1), \ldots, y_n(T_n))$. This yields

$$D(\theta|\mathcal{M}) = -2 \sum_{n=1}^{N} \sum_{t=1}^{T_n} y_n(t)\lambda(t|x_n) + (1 - y_n(t))(1 - \lambda(t|x_n))$$

with $\lambda(t|x_n) = \exp(\eta_t)/1 - \exp(\eta_t)$ giving the conditional probability of failure at time t. Here for all individuals the sum is taken over all time points up to its last observation.

For model selection procedures, the posterior expected deviance is known to prefer models which are overly complex. Therefore model dimension, i.e. the effective number of model parameters, must be involved into the selection process. To be more specific, the goodness of fit measured by $\overline{D}(\mathcal{M})$ must be adjusted by some penalty term depending on model dimension. For a parametric model this is an easy task, as the model dimension is known to be the number of unknown parameters. However, specifying the dimension of a non-parametric model is not obvious. Details are given in the subsequent section, where a Bayesian penalised model criteria is described together with a general estimate of model dimension.

4.4.5 The Deviance Information Criterion

In a recent research report Spiegelhalter et al. (2001) present a penalised model criterion for comparison of models of arbitrary dimension together with a proposal for the determination of the complexity of a model. Based on Dempster's suggestion to consider the posterior distribution of the deviance, they introduce the *Bayesian Deviance Information criterion* (DIC)

$$\text{DIC}(\mathcal{M}) = \overline{D}(\mathcal{M}) + df_{\mathcal{M}}. \tag{4.33}$$

The DIC can be viewed as a Bayesian version of the classical information criterion where the goodness of fit of a model is measured by the posterior mean of the deviance which is penalised by the effective model dimension $df_{\mathcal{M}}$.

The potency of this proposal lies in the definition of a measure for model dimension, since, as mentioned above, in non-parametric modelling the effective number of parameters is usually unclear. Spiegelhalter et al. cope with this problem by defining the effective number of the parameters $df_{\mathcal{M}}$ as the difference between the posterior mean of the deviance and the deviance of the posterior means of the parameters, that is

$$
\begin{aligned}
df_{\mathcal{M}} &= \text{E}_{\theta|y}[D(\theta)] - D(\text{E}_{\theta|y}[\theta]) \\
&= \overline{D}(\mathcal{M}) - D(\bar{\theta}|\mathcal{M}).
\end{aligned}
\tag{4.34}
$$

It is immediately seen that $df_{\mathcal{M}}$ is subject not only to the definition of model $\mathcal{M}$ but is also influenced by the data y. Moreover, by involving $D(\bar{\theta}|\mathcal{M})$, it depends on the parameter in focus, that is the parameter, from which the posterior mean $\bar{\theta}$ is calculated. In the context of regression models when variable selection or decision on effect structures are of interest, the parameter in focus θ usually comprises the covariate effects. This is of course not mandatory. When comparing survival models of different classes one may, for example, alternatively choose the hazard rate to be the parameter of interest. Considering a univariate logit model for discrete survival data one would then regard $D\big(\frac{1}{G}\sum f(\beta^{(g)})\big)$, with $f(\beta) = \exp(\beta x)/1 - \exp(\beta x)$, instead of $D\big(\frac{1}{G}\sum \beta^{(g)}\big)$. This will usually yield different values for the estimated model dimension. Spiegelhalter et al. argue that their measure of complexity should reflect 'difficulty in estimation'. Therefore it is reasonable that beside the prior information of the model it also depends on the parameter in focus as well as on the actually observed data. In the case of dynamic modelling this would mean that a covariate effect following a bent course will induce a larger $df_{\mathcal{M}}$ than a covariate effect that shows linear or no time-variation, even if both models are using the same prior assumptions and in particular the same smoothing prior. This suits to the fact, that Bayesian dynamic logit modelling presented in Section 4.2 includes a data driven determination of the smoothing parameter, i.e. the random walk variance. Note that this data-dependency of the $df_{\mathcal{M}}$ is in particular contrast to a non-Bayesian view point, where the degree of a model is usually not related to the data at all. In non-Bayesian dynamic modelling the degree of model complexity is fixed by the chosen smoothing parameter.

To justify the definition of $df_{\mathcal{M}}$ given in (4.34), Spiegelhalter et al. consider the definition of the residual information in the data y conditional on the parameter θ, that is $-2\log l(y|\theta)$ and regard an estimate $\tilde{\theta}$ for the unknown parameters. The authors argue that the difference

$$U(y, \theta, \tilde{\theta}) = -2\log l(y|\theta) + 2\log l(y|\tilde{\theta}) \qquad (4.35)$$

expresses "the reduction in surprise or uncertainty due to estimation, or alternatively the degree of over-fitting due to $\tilde{\theta}$ adapting to the data", (Spiegelhalter et al. 2001, p. 3). It can therefore be used to form a measure of model dimensionality. Within the Bayesian framework, rather than an 'estimate', the posterior mean of the parameters $\tilde{\theta} = \mathrm{E}_{\theta|y}[\theta] = \bar{\theta}$ is used to construct a model. Taking posterior expectation of $U(y, \theta, \bar{\theta})$ leads to

$$
\begin{aligned}
\mathrm{E}_{\theta|y}[U(y, \theta, \bar{\theta})] &= \mathrm{E}_{\theta|y}[-2\log l(y|\theta)] + 2\log l(y|\bar{\theta}) \\
&= \overline{D} - D(\bar{\theta}) \\
&=: df_{\mathcal{M}}
\end{aligned}
$$

which is the proposed definition of the effective number of the parameters. The determination of the model dimension by means of $df_{\mathcal{M}}$ is straightforward using the MCMC output. Moreover due to Jensen's inequality, that is $f(\bar{x}) < \overline{f(x)}$ for any concave function f, $df_{\mathcal{M}}$ is certain to be positive as long as the log likelihood is concave in θ.

The measure of model dimension $df_{\mathcal{M}}$ now allows to transform the DIC($\mathcal{M}$) into

$$
\begin{aligned}
\text{DIC}(\mathcal{M}) &= \overline{D}(\mathcal{M}) + df_{\mathcal{M}} \\
&= 2\overline{D}(\mathcal{M}) - D(\bar{\theta}|\mathcal{M}) \\
&= D(\bar{\theta}|\mathcal{M}) + 2df_{\mathcal{M}}.
\end{aligned}
\tag{4.36}
$$

With (4.36) the DIC has the common structure of the Akaike Information criterion (Akaike, 1973). The DIC penalises this plug-in measure of fit by twice the effective number of parameters.

Spiegelhalter et al. consider their model selection criteria DIC as a 'semi-formal' approach, as it has no decision-theoretic foundation. They do, however, describe an approximate decision-theoretic justification for the DIC involving loss functions and replicate data. For ease of notation I omit the model index $\mathcal{M}$ in the following. Let $\mathcal{L}(y, \tilde{\theta})$ denote the loss when data y is described by model $l(y|\tilde{\theta})$ using an estimate $\tilde{\theta} = \tilde{\theta}(y_{obs})$. The loss $\mathcal{L}(y_{obs}, \tilde{\theta})$ is considered to be the 'apparent loss', that is the loss induced by the actual observations. The 'optimism' associated with the estimate $\tilde{\theta}$ is then expressed as

$$
c(y, \theta, \tilde{\theta}) = E_{y_{rep}|\theta}[\mathcal{L}(y_{rep}, \tilde{\theta})] - \mathcal{L}(y_{obs}, \tilde{\theta}),
$$

that is the expected loss when observing additional data y_{rep} adjusted by the apparent loss. For the logarithmic loss function $\mathcal{L}(y, \tilde{\theta}) = -2\log l(y|\tilde{\theta}) = D(\tilde{\theta})$ this 'optimism' is

$$
c(y, \theta, \tilde{\theta}) = E_{y_{rep}|\theta}[D_{rep}(\tilde{\theta})] - D(\tilde{\theta}).
\tag{4.37}
$$

The authors show, that from (4.37) the posterior expectation of the 'optimism' is asymptotically equal to twice the effective numbers of parameters. This means that the DIC approximately represents the expected posterior loss $E_{\theta|y}\left[E_{y_{rep}|\theta}[D_{rep}(\tilde{\theta})]\right]$ when adapting a model $\mathcal{M}$ to data y.

Further asymptotic properties of the DIC and $df_{\mathcal{M}}$ are derived from the Bayesian central limit theorem (4.5), i.e. for the large sample situation of an approximate normal likelihood and negligible prior information. Let $L_\theta = \log l(y|\theta)$ denote the log-likelihood so that $D(\theta) = -2L_\theta$. The first and second derivatives L_θ' and L_θ'' give the score function and minus the observed Fisher information. Further, let $\hat{\theta}$ denote the maximum likelihood estimate of the p-dimensional parameter vector with $L'_{\hat{\theta}} = 0$. The Bayesian central limit theorem states that for large samples the posterior probability is approximately normal with

$$\theta|y \overset{a}{\sim} N(\hat{\theta}, [-L''_{\hat{\theta}}]^{-1}). \tag{4.38}$$

With (4.38) the posterior mean is asymptotically equal to the maximum likelihood estimate, that is $\bar{\theta} = \hat{\theta}$, and the posterior variance is approximately the inverse Fisher information evaluated at the maximum likelihood estimate. Expanding the deviance $D(\theta)$ around $\hat{\theta}$ gives

$$D(\theta) \approx D(\hat{\theta}) + (\theta - \hat{\theta})' \frac{\partial D}{\partial \theta}\bigg|_{\hat{\theta}} + \frac{1}{2}(\theta - \hat{\theta})' \frac{\partial^2 D}{\partial \theta^2}\bigg|_{\hat{\theta}} (\theta - \hat{\theta})$$

$$= D(\hat{\theta}) + 2(\theta - \hat{\theta})' L'_{\hat{\theta}} - (\theta - \hat{\theta})' L''_{\hat{\theta}}(\theta - \hat{\theta})$$

Since $L'_{\hat{\theta}} = 0$ it follows that

$$D(\theta) - D(\hat{\theta}) \approx -(\theta - \hat{\theta})' L''_{\hat{\theta}}(\theta - \hat{\theta}) \tag{4.39}$$

where with (4.38) the right term of (4.39) is approximately χ^2_p-distributed. Taking posterior expectation of (4.39) and recalling that under asymptotic normality the posterior mean is equal to $\hat{\theta}$, one gets

$$df_{\mathcal{M}} := E_{\theta|y}[D(\theta)] - D(\hat{\theta}) \approx p. \tag{4.40}$$

Thus, $df_{\mathcal{M}}$ is asymptotically equal to the true number of parameters p when approximately normal likelihoods and negligible prior information are assumed. It becomes clear that in this situation the DIC is asymptotically equivalent to the Akaike information criterion AIC $= D(\hat{\theta}) + 2p$ (Akaike, 1973).

The deviance-based approach suggested by Spiegelhalter et al. has some especially appealing characteristics. Most important is its general applicability to arbitrary complex models, since both, the measure of fit and the measure of model complexity are derived from the model deviance and its posterior expectation. These may easily be obtained based on the posterior sample resulting from an Markov Chain Monte Carlo run for model estimation, so that as soon as a model-posterior is generated, computation of the DIC is straightforward and does not require any relevant additional effort. Another appealing characteristic of the approach is that the DIC has the common form of an information criterion, constructed by a measure of fit which is penalised by model complexity and can be viewed as the Bayesian analogue to the AIC. Additionally, the appendant measure of model dimension may give useful insights into the general model complexity and possible dependencies between the parameters.

Spiegelhalter et al. also provide an interpretation scheme for the deviance information criterion, which can be used for model rating. It is rather rough and follows Jeffreys guidelines for the Bayes factor although the DIC usually will yield different values. Regarding the difference in the DIC between model $\mathcal{M}_0$ and $\mathcal{M}_1$, denoted by $\Delta\text{DIC}_{(0,1)} = \text{DIC}(\mathcal{M}_0) - \text{DIC}(\mathcal{M}_1)$, this scheme can be resumed by stating that for $0 < \Delta\text{DIC}_{(0,1)} \leq 2$ both models can be considered as fitting equally well, while a difference $2 < \Delta\text{DIC}_{(0,1)} < 7$ already shows a clear superiority of model $\mathcal{M}_1$.

4.5 A Simulation Study

In the following the performance of the different Bayesian model criteria are examined in a comparative simulation study. The main focus is thereby on the comparison of logit models with time-constant and time-varying effects in different situations. Thereto 100 samples of failure-time data are generated from a logistic setting with five different dynamic effect structures. Similar to the simulation study in Chapter 3 all samples consist again of two groups, a 'baseline group' with $x = 0$ and a 'risk group' with $x = 1$, where the two groups have equal size.

In the first three scenarios the hazard rate for an event is chosen to be generally rather high with a baseline hazard of $\lambda_0 = e^{-2}/(1 + e^{-2}) = 0.12$. This produces observation times that are rather short with a high rate of equal failure times, i.e. many ties. The permanent probability for censoring was chosen to be $P_c = 0.05$. Hence, data are generated for which discrete survival models are most suitable. For each group ($x = 0$ and $x = 1$) 100 observations are simulated where the following effect structures are used:

(1) A constant setting with: $\beta(t) = 1$, to investigate how consistent the choices based on the different criteria are.

(2) A linear time-dependency: $\beta(t) = -0.15t + 1$, describing an effect, which declines over time.

(3) A steep quadratic function: $\beta(t) = -3 + 1.2t - 0.06t^2$, producing an effect that raises in the first 12 time-point and decreases afterwards (quadratic I).

In the following scenarios the event probabilities are reduced considering a baseline hazard of $\lambda_0 = e^{-4}/(1 + e^{-4}) = 0.018$ as in Chapter 3, which produces longer observation times. The probability of censoring was scaled down to $P_c = 0.005$. Data are generated with two dynamic effect structures:

(4) A cosinus function: $\beta(t) = 1.25\cos(0.075t)$, with 200 observations in each group. This simulation setting is used to investigate the performance in more complex dynamic situations.

(5) A flat quadratic function: $\beta(t) = -1 + 0.08t - 0.0008t^2$, with 100 observations per group. To investigate the impact of priors on the criteria, the data from this setting was fitted twice with different prior specification, (quadratic II and quadratic III).

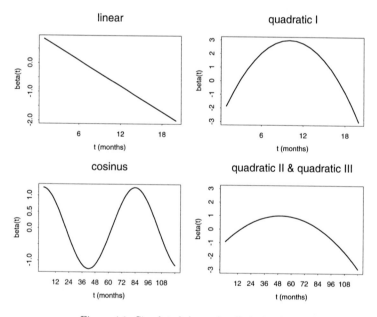

Figure 4.1: Simulated dynamic effect structures.

Figure 4.1 gives the functions used for the four different dynamic effect functions. For fitting Bayesian logit models with dynamic and constant covariate effects the hybrid Markov Chain Monte Carlo algorithm is used as described in Section 4.3. For the second order random walk variance a diffuse prior $IG(0.1, 0.001)$ was chosen for all five settings. To investigate the impact of prior specifications on the criteria, the samples of scenario 5 are re-fitted using an informative (non-vague) prior $IG(10, 0.001)$ that however need not to meet the true value. A MCMC sampling procedure was implemented in C++ under Unix, which efficiently handles discrete survival data so that computation time of the simulations could be noticeably reduced. Block sizes range from 6 to 12 and the normal proposal function of the constant effects had an variance of 0.025. This assured an overall acceptance rate of about 50 %. For each simulated sample a model with a constant covariate effect and a model with a dynamic covariate effect was fit by generating a Markov chain of 15000 samples, after a burn-in of 5000 draws. To reduce autocorrelation of the sample used for inference, it was thinned out by using only every 10th element of the chain.

For each simulated sample the Bayes factor estimated by the harmonic mean, the posterior Bayes factor, the posterior mean of the deviance $\overline{D}$, the deviance of the posterior model $D(\overline{\theta})$ and the Deviance Information criterion are calculated for the constant model $\mathcal{M}_0$ and the dynamic model $\mathcal{M}_1$. To decide on dynamic effects the criteria or the differences of the criteria, respectively, are evaluated, i.e. $2log\mathrm{BF}_{1,0} = 2\log P(y|\mathcal{M}_1) - 2\log P(y|\mathcal{M}_0)$, $2log\mathrm{PBF}_{1,0} = 2\log \overline{l}(y|\mathcal{M}_1) - 2\log \overline{l}(y|\mathcal{M}_0)$, $\Delta\overline{D}_{0,1} = \overline{D}(\mathcal{M}_0) - \overline{D}(\mathcal{M}_1)$, $\Delta D_{0,1} = D(\overline{\theta}, \mathcal{M}_0) - D(\overline{\theta}, \mathcal{M}_1)$ and $\Delta\mathrm{DIC}_{0,1} = \mathrm{DIC}(\mathcal{M}_0) - \mathrm{DIC}(\mathcal{M}_1)$. Positive values of these differences represent a clear evidence for a dynamic covariate effect, while negative values indicate, that the constant model is preferred.

Figure 4.2 shows the empirical densities of the differences for the six settings resulting from the simulated samples and Table 4.3 gives some quantiles of the distributions. In Figure 4.2 vertical lines mark the classification bounds of the interpretation scheme for the Bayes factor given in Table 4.1, where differences > 2 indicate a positive evidence for the dynamic model, > 6 indicate strong evidence and > 10 indicate very strong evidence, and vice versa. (For more clearness of the plots, the empirical distribution of $\Delta D(\overline{\theta})$ is not displayed. It is generally known that $D(\overline{\theta})$ tends to decrease with increasing complexity of the model, which is clearly seen in Table 4.3.)

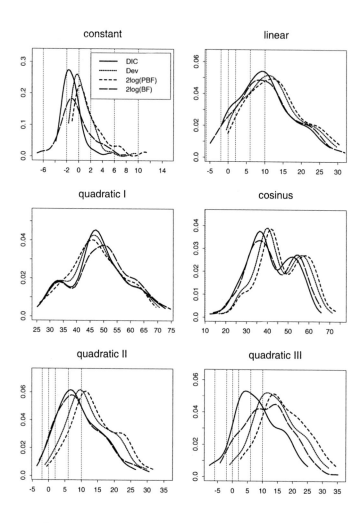

Figure 4.2: Distribution of the differences in the criteria of the constant fit $\mathcal{M}_0$ and the dynamic fit $\mathcal{M}_1$.

Table 4.3: Distribution of the differences in the criteria of the constant fit $\mathcal{M}_0$ and the dynamic fit $\mathcal{M}_1$

	criterion	min	median	max	percentage < 2	< 6	< 10
constant	$2\log \mathrm{BF}_{1,0}$	-7.02	-0.91	7.93	81	99	100
	$2\log \mathrm{PBF}_{1,0}$	-1.10	1.01	11.47	64	90	97
	$\Delta \overline{D}_{0,1}$	-1.68	0.26	9.60	78	98	100
	$\Delta D_{0,1}$	-0.06	1.91	13.74	53	93	98
	$\Delta \mathrm{DIC}_{0,1}$	-3.85	-1.42	5.46	95	100	100
linear	$2\log \mathrm{BF}_{1,0}$	-4.89	9.84	31.98	16	29	50
	$2\log \mathrm{PBF}_{1,0}$	-0.08	12.24	30.85	5	22	39
	$\Delta \overline{D}_{0,1}$	-0.54	11.23	29.74	5	23	42
	$\Delta D_{0,1}$	0.68	12.81	31.59	3	16	33
	$\Delta \mathrm{DIC}_{0,1}$	-1.77	9.85	27.94	17	31	50
quadratic I	$2\log \mathrm{BF}_{1,0}$	25.21	49.72	74.76	0	0	0
	$2\log \mathrm{PBF}_{1,0}$	26.18	46.78	69.35	0	0	0
	$\Delta \overline{D}_{0,1}$	25.69	47.03	70.58	0	0	0
	$\Delta D_{0,1}$	25.23	46.34	68.49	0	0	0
	$\Delta \mathrm{DIC}_{0,1}$	26.14	47.88	72.74	0	0	0
cosinus	$2\log \mathrm{BF}_{1,0}$	12.42	40.21	65.78	0	0	0
	$2\log \mathrm{PBF}_{1,0}$	17.92	45.71	72.68	0	0	0
	$\Delta \overline{D}_{0,1}$	16.74	44.27	71.18	0	0	0
	$\Delta D_{0,1}$	19.49	48.21	76.46	0	0	0
	$\Delta \mathrm{DIC}_{0,1}$	14.00	39.98	65.91	0	0	0
quadratic II	$2\log \mathrm{BF}_{1,0}$	-3.56	8.39	30.56	14	34	61
	$2\log \mathrm{PBF}_{1,0}$	-0.58	12.75	32.10	3	12	28
	$\Delta \overline{D}_{0,1}$	-1.11	10.76	30.18	4	16	40
	$\Delta D_{0,1}$	0.65	13.87	33.86	2	9	19
	$\Delta \mathrm{DIC}_{0,1}$	-2.87	8.05	26.51	11	31	61
quadratic III	$2\log \mathrm{BF}_{1,0}$	-4.13	12.36	35.16	13	23	45
	$2\log \mathrm{PBF}_{1,0}$	2.69	16.58	34.40	0	7	17
	$\Delta \overline{D}_{0,1}$	-0.86	14.02	32.11	5	9	28
	$\Delta D_{0,1}$	5.86	20.46	39.27	0	1	7
	$\Delta \mathrm{DIC}_{0,1}$	-7.66	7.72	25.08	19	44	63

In the constant setting all densities have their mode below 2. The DIC penalises complexity the strongest. It yields differences smaller than zero for 87 % of the samples and shows a positive evidence for the constant model $(\Delta \text{DIC} < -2)$ for 30 %. Also the Bayes factor finds positive evidence for the constant model in 28 % and has negative differences $(2 \log \text{BF}_{1,0} < 0)$ for 63 %. The density of the posterior mean of the deviance $\Delta \overline{D}_{0,1}$ is spread around zero with 44 % of the values below zero. The posterior Bayes factor in general yields larger differences; only in a quarter of all cases $2 \log \text{PBF}_{1,0}$ is smaller zero and 36 % lie even above 2. This supports Aitkin's observation that the posterior Bayes factor does not fit into the interpretation scheme for the Bayes factor given in Table 4.1, (Aitkin, 1997).

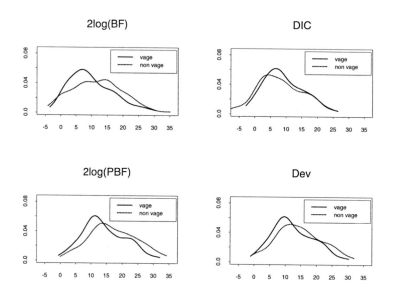

Figure 4.3: Distribution of the differences in the criteria of the constant fit $\mathcal{M}_0$ and the dynamic fit $\mathcal{M}_1$ for different priors.

In the linear setting the ordering of the distributions is generally the same, where the DIC yields smaller differences while $2 \log \text{PBF}_{1,0}$ yields larger values and the values of $2 \log \text{BF}_{1,0}$ are more spread. However, for most data sets all criteria favour the varying model.

In the setting 'quadratic I' the dynamic structure is absolutely obvious and all criteria select the dynamic model in all cases. The densities of the differences lie one upon the other. Also in the cosinus setting there is no doubt about time-varying effects and all criteria are in agreement in all cases. In the quadratic II setting, however, some disagreement can be seen. The Bayes factor and the DIC find no notable positive evidence for the dynamic model in 14 % and 11 %, while $\Delta \overline{D}_{0,1}$ and $2 \log \text{PBF}_{1,0}$ are < 2 in only 4 % and 3 %. When an informative prior $IG(10, 0.001)$ is used for the variance of the second order random walk (quadratic III setting), the densities change noticeably. This is illustrated in more detail in Figure 4.3, where for each criterion the density resulting from the quadratic II setting based on a flat prior $IG(0.1, 0.001)$ is compared to the density resulting from the quadratic III setting using the prior $IG(10, 0.001)$. The densities resulting from the latter setting are generally more spread. This discrepancy is the largest for the Bayes factor, where for the quadratic III setting the distribution of $2 \log \text{BF}_{1,0}$ is very flat and shifted towards high values. This means that in this setting the Bayes factor stronger supports the dynamic fit $\mathcal{M}_1$ than in does the quadratic II setting. The Deviance Information criterion seems to be quite robust towards these changes in the priors. The sensitivity of the Bayes factor to the prior specifications has been noted above and will be regarded in more detail in Chapter 5.

With the introduction of the Deviance Information criterion Spiegelhalter et al. give an estimate for the effective number of parameters, which allows to evaluate the complexity of the constant fit $df(\mathcal{M}_0)$ and of the dynamic fit $df(\mathcal{M}_1)$. The difference $\Delta df_{1,0} = df(\mathcal{M}_1) - df(\mathcal{M}_0)$ can be seen as the additional number of parameters needed when allowing for time-variation in the covariate effects. Comparison of the effective number of parameters of different models provides an informative insight into dependency structures between effects and is generally useful in any practical analysis. Table 4.4 lists median, minimum and maximum for both, the effective number of parameters of the models as well as for the differences $\Delta df_{1,0}$. Figure 4.4 displays the empirical densities of $\Delta df_{1,0}$.

Table 4.4: Distribution of the 'effective number of parameters' of the constant fit $\mathcal{M}_0$ and the dynamic fit $\mathcal{M}_1$ and their differences Δdf

		min	median	max
constant	$df(\mathcal{M}_0)$	3.08	3.66	6.58
	$df(\mathcal{M}_1)$	4.18	5.43	9.95
	$\Delta df_{1,0}$	0.32	1.73	4.43
linear	$df(\mathcal{M}_0)$	3.04	3.73	6.09
	$df(\mathcal{M}_1)$	4.40	5.43	8.29
	$\Delta df_{1,0}$	0.33	1.75	4.66
quadratic I	$df(\mathcal{M}_0)$	5.27	6.42	8.39
	$df(\mathcal{M}_1)$	3.91	5.55	8.72
	$\Delta df_{1,0}$	−3.17	−0.82	2.89
cosinus	$df(\mathcal{M}_0)$	5.59	6.67	9.45
	$df(\mathcal{M}_1)$	8.91	11.00	12.99
	$\Delta df_{1,0}$	0.61	4.23	6.62
quadratic II	$df(\mathcal{M}_0)$	4.28	4.99	6.28
	$df(\mathcal{M}_1)$	6.42	8.25	10.89
	$\Delta df_{1,0}$	1.76	3.19	4.81
quadratic III	$df(\mathcal{M}_0)$	9.29	10.56	13.95
	$df(\mathcal{M}_1)$	15.28	16.94	20.12
	$\Delta df_{1,0}$	3.00	6.31	9.14

The data from the constant setting and from the linear setting require the same number of parameters, where the dynamic covariate effect in model $\mathcal{M}_1$ increases complexity by about 1.75. For the constant setting this increase of complexity might seem somewhat surprising. The estimate of effective number of parameters $df = \overline{D} - D(\bar{\theta})$ clearly depends on the data, and one would not expect the dynamic model to be more complex if the data does not contain any dynamic structure. The fact that the dynamic model for the constant setting requires as many parameters as the dynamic model for the linear setting can be explained by the use of a second order random walk to describe the temporal development of the effects. It penalises deviations of the straight line and hence does not distinguish between a constant and a linear effect structure.

Fitting the data generated from the cosinus setting generally requires more parameters compared to the models for data from the constant setting. This is seen for the dynamic model $(df_{med}(\mathcal{M}_1) = 11)$ as well as for the model with a constant effect $(df_{med}(\mathcal{M}_0) = 6.67)$. This derives from the fact that in all our models the baseline effect is included in a dynamic fashion (see Section 2.3). It therefore partly takes on dynamic structure of the data when the covariate effect is not allowed to vary with time.

The data of the quadratic II setting requires a median complexity of about five for the constant model $\mathcal{M}_0$ which increases to $df_{med}(\mathcal{M}_1) = 8.25$ when a dynamic covariate effect is included. The median of the differences is 3.19.

When the informative prior $IG(10, 0.001)$ is used to fit the data (quadratic III setting) the complexity strongly raises for both models, $\mathcal{M}_0$ and $\mathcal{M}_1$. Note that the constant model is affected by the prior through the dynamic baseline effect. The median number of parameters of the constant model is $df_{med}(\mathcal{M}_0) = 10.56$. In the dynamic model the median complexity is nearly 17. The high increase in complexity between the constant and the dynamic fit in this setting $(\Delta df_{1,0} = 6.31)$ is cause of the rather low values in $\Delta \mathrm{DIC}_{0,1}$ (see Figure 4.2 and Figure 4.3).

The most astonishing results are observed for the quadratic I setting, where the number of effective parameters of the dynamic model is smaller than the number of parameters of the constant model. This would mean, that the complexity is reduced by allowing the covariate effect to vary with time, which seems illogical.

In particular as the criteria themselves all clearly favour the dynamic fit. (It should be noted, that the convergence of the MCMC algorithm in this setting was as good as for the other settings.)

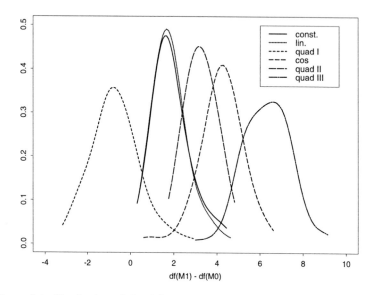

Figure 4.4: Distribution of the differences in the effective number of parameters Δdf.

The simulation study showed the behaviour of the Bayes factor, the posterior Bayes factor, the mean Deviance and the Deviance Information criterion in different situation of varying complexity. The Bayes factor generally shows a larger variation in the evaluation of the dynamic effect than the other criteria, i.e. for all settings in Figure 4.2 the densities of $2 \log \mathrm{BF}$ are wider and flatter. The posterior Bayes factor generally puts more value on the more complex models. For the Deviance Information criterion per definition the differences $\Delta \mathrm{DIC}_{0,1}$ are generally smaller than $\Delta \overline{D}_{0,1}$, given that the effective number of parameters increases. In the given simulations, the DIC penalises model complexity strongest, which is in contrast to what we would expect from theoretical large sample approximations as considered in Chapter 5.

The investigation of the impact of prior assumption showed the expected results; while the Deviance Information criterion is rather robust towards prior specifications, the Bayes factor clearly is influenced the most. Note that beside the vague prior of the random walk variance also improper priors have been used for the constant effect and the initial values of the dynamic effects. Yet, these improper priors cause no computational problems if the Bayes factor is approximated by the posterior harmonic mean. The impact of prior specifications on the Bayes factor is discussed in more detail in Chapter 5.

The analysis of the behaviour of the effective number of parameters give an interesting insight into the functionality of the DIC-concept. When comparing the estimated complexity of the constant models from the different settings, it is seen that models generally become more complex (in terms of df) when the structure of the data is more complex: the number of parameters of the constant models are clearly higher in the quadratic and in the cosinus setting than in the constant setting. In the linear setting the same number of parameters is found as in the constant setting, reflecting that the second order random walk can not distinguish between a constant and a linear effect structure. Linear effects are therefore harder to identify within this model framework. This is reflected in the low values of differences of the criteria in Figure 4.2. It is also seen that in the constant setting the dynamic model requires more parameters than the constant model, although the data does not contain any dynamic covariate effect structures. This could indicate that each dynamic effect requires a certain minimal number of parameters, independent of the complexity of the data. On the other hand this is in contradiction to the fact that the number of parameters decreases in the quadratic I setting. This observation that the estimate of the effective number of degrees of freedom $df = \overline{D} - D(\overline{\theta})$ sometimes decrease although the complexity of the model predictor formally is increased has been made in other applications before. For instance it can be seen in the lip cancer example of the original paper on the Deviance Information criterion (Spiegelhalter et al., 2001) where a model with an exchangeable random effect yields a smaller number of effective parameters than a model that additionally includes spatial effects. However the authors do not refer to this phenomenon.

Generally, the simulation study reveals that model evaluations by the Bayes factor, the posterior Bayes factor, the mean Deviance and the Deviance Information criterion do not substantially differ when used to examine dynamic effect structures. Especially if if for model selection coarse interpretation schemes are applied, such as listed in Table 4.1, the differences between the criteria clearly lose relevance.

4.6   Infant Mortality in Zambia

Mortality, in particular child mortality, is among the most important measures of well-being and development in poor countries. Socio-economic analyses of child mortality involve the education of the parents, the income or wealth situation of the household, access to water and sanitation services, access to health services, etc.. In this study I will explore the formation of infant mortality in Zambia where, additionally to socio-economic factors, maternal factors such as birth spacing or maternal age at birth and nutrition-related issues are considered. The major objective is again to assess possible time-variation in the effects of the influential factors. This arises from the belief that some determinants of child mortality, such as the impact of the preceding birth interval, might have a larger impact in the early phase of life. Also, breastfeeding is likely to have a different impact on mortality in different life periods. In particular, while there is strong biomedical evidence that exclusive breastfeeding in the first months of life lowers child mortality (WHO, 1995), the importance of breastfeeding is less clear after the first year of life. Similarly, the mother's education might have a different impact on child mortality in different stages of life.

The analysis is based on the 1992 Demographic and Health Survey (DHS) carried out by MEASURE DHS+ program. For a representative sample of women from Zambia data were collected on the monthly survival times of their children. Additional covariates which might influence the risk of mortality were recorded. In previous studies on infant mortality in developing countries, survival times were often grouped into rather coarse classes, such as death before or after 6 months. A static logit or probit framework was then used, thus modelling the probability of being alive at a certain age. While this is partly driven by the lack of time-varying covariates, it is nevertheless problematic as it fails to model the precise timing of death and thus ignores information available in most cross-sectional micro data sets. Moreover, many data sets, including the DHS used here, collect a large number of retrospective information for each child. Here for a refined analysis of the causes of infant mortality discrete-time survival models are used as described in Section 4.2, using months as time units.

To compare models with time-varying and time-constant effects, that is to decide which effects show distinct dynamic structure, the different model diagnostic tools of Section 4.4 are employed, including the Bayes factor estimated by the posterior harmonic mean, the posterior Bayes factor, the posterior deviance, the deviance of the posterior mean model and finally the DIC together with the effective number of degrees of freedom.

4.6.1 The Data

The data are from the 1992 household survey of the Demographic and Health Surveys (DHS) for Zambia. Zambia is a low income developing country in Southern Africa, belonging to the ten poorest countries in the world. While it was one of the richer African countries in the 1960s, largely due to its great mineral wealth (especially copper), the economy stagnated and regressed throughout most of the 1970s and 1980s. As a result, poverty in Zambia has increased considerably (by 1996, 79 % of the population was living on less than US$1 per capita per day) and with it infant mortality rose throughout the 1980s and early 1990s (World Bank, 1995).

The DHS program is funded by United States Agency for International Development to collect population, health, and nutrition data from developing countries and is implemented by Macro International Inc. For the survey a nationally representative sample of women in reproductive ages (15-49) was interviewed using a household questionnaire and a women's questionnaire. The questionnaires consist of different sections including respondent basic data such as age, educational achievements and woman's work status, the district type and condition of residence, data on reproduction and birth history with each child's individual characteristics and health history, information on maternity and feeding of the child, among others. (For more details on the data see http://www.measuredhs.com/.) In order to undertake survival modelling child individual records were constructed from the DHS data sets, that consist of the child's survival information and a set of the explanatory covariates.

The general medical definition distinguishes mortality of a child with respect to the child's age assuming different causes: While death within the first week of life is classified as *perinatal mortality* and death within the first month is referred to as *neonatal mortality, infant mortality* specifies the death occurring after the first month.

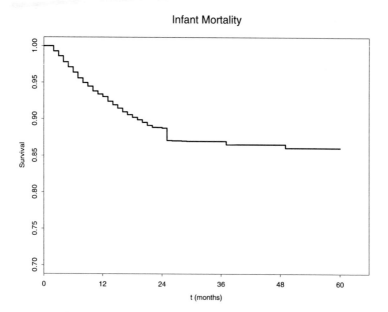

Figure 4.5: Overall survival of children less than five years of age in Zambia.

For our analysis individual data records were available for 5965 children that were born within the last five years before interview and had survived their first month of life. 633 children had died. The survival curve given in Figure 4.5 shows a 5-year mortality rate for children of Zambia of nearly 14 %. Moreover, it reveals that most of the events occur within the first two years. (Note that here, as well as in the following KM-plots, the y-axis is scaled to $[0.7, 1]$.) For comparison, the corresponding 5-year infant mortality rate of Germany for 1992 is under 0.8 % (Statistisches Bundesamt, 1994).

Since survival time is recorded in months and total observation time is restricted to 60 months, the data naturally contains a large number of tied events. Because of this, a logit model for discrete survival data seems more suited to fit the data rather than a survival model for continuous time.

4.6.2 Coding of the Covariates

To analyse dynamic effect structures a set of covariates was constructed from the basic variables of the DHS data which are grouped to form categorical and binary coded factors. Table 4.5 gives a short description of these constructed covariates and their distribution. Child and birth specific information is given by the covariates gender, birth order and the preceding birth interval, where birth order is reduced to a binary factor, indicating whether the child is the first born or not. Duration of breastfeeding is given in months. The preceding birth interval is calculated as months between the birth of the child and the mother's previous birth and is naturally only defined for children who are not first borns. A 'short' birth interval is defined to be no longer than two years. Maternal information includes the age of the mother at birth, calculated from the DHS data on the date of birth of the mother and the date of birth of the child. Also the highest educational attainment of the mother is considered, distinguishing no education or incomplete/complete primary school and higher than primary school. The indicator whether the mother is currently working and the total number of persons currently living in the household might provide information about the child's care and in particular might mirror the wealth of the household. Information on the type and condition of the residence comprises the type of district by distinguishing rural and urban, whether the household has electricity and the main material of the floor. The major source of drinking water is divided with respect to its quality, where water access in the residence (including bottled water) or from public tab is assumed have controlled quality, while water from public wells, springs, rivers or streams, ponds or lakes or rainwater is not controlled. Water from tanker trucks is also added to the latter category since, even it is controlled, it is usually rather costly and scarce.

Some covariates naturally have special characteristics which must be taken into account when fitting the data. This applies for breastfeeding, which is an internal covariate that is observed only so long as the child survives and is uncensored. In consequence, the covariate 'duration of breastfeeding' carries survival information of the corresponding child as it can never exceed its survival time. Therefore this covariate can not be included into the model as if its value was fixed at birth but must be handled differently. Instead of using duration of breastfeeding in months, I generated a binary covariate process, which is equal 1 during the months the child was breastfed and 0 when the child was fed differently.

Table 4.5: Factors analysed in the infant mortality study

factor	frequency	coding	interpretation
SEX	3003 (50.3 %)	0: female	gender
	2962 (49.7 %)	1: male	
BFIRST	4669 (78.3 %)	0: no	whether the child
	1296 (21.7 %)	1: yes	is first born
BIRTHIN*	1035 (22.2 %)	0: $\leq$ 24 months	preceding
	3628 (77.8 %)	1: > 24 months	birth interval
MAGE	1780 (28.8 %)	0: $\leq$ 21	age of mother
	3486 (58.4 %)	1: 22-35	at birth
	699 (11.7 %)	2: > 35	
MEDUCATION*	4822 (80.9 %)	0: $\leq$ primary	mother's educational
	1141 (19.1 %)	1: > primary	attainment
MWORKING*	2708 (45.4 %)	0: no	mother
	3255 (54.6 %)	1: yes	currently working
HHSIZE	1177 (19.7 %)	0: 1 - 4	total number
	1522 (25.5 %)	1: 5 - 6	of household members
	3266 (54.8 %)	2: 7 -	
URBAN	3393 (56.9 %)	0: rural	type of district
	2572 (43.1 %)	1: urban	of residence
WATER*	2835 (47.7 %)	0: residence/tap	source of
	3108 (52.3 %)	1: else	water
HOUSE*	3104 (52.3 %)	0: wood/sand	material
	2836 (47.7 %)	1: higher quality	of the floor
ELECTRICITY*	4776 (80.3 %)	0: no	electricity
	1170 (19.7 %)	1: yes	
BREASTF*		0: no	currently breastfeeding
		1: yes	(time-dependent)

* = contains missing values

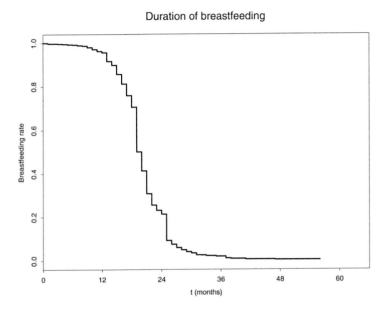

Figure 4.6: Duration of breastfeeding (corrected for break-off by death or censoring of the child).

To give an example, consider a child who survived only five months but was breastfed during all this time. Duration of breastfeeding is then equal to five. The corresponding covariate process is equal one for the first five month and not defined later. In contrast, if a mother stopped breastfeeding after five months for some other reasons, e.g. illness, duration of breastfeeding is equal to five as well, while the covariate process equals one for five months and equals zero for the following months until end of observation.

The hazard function for internal time-dependent covariates is defined such that it conditions on the covariate value at the current time t, as described in Chapter 2. With this convention parameter estimation of the logit model for internal covariates is identical to external time-dependent covariates. Figure 4.6 gives an impression of the distribution of the duration of breastfeeding within the sample. Here, an observation is censored, when duration of breastfeeding ended due to death or censoring. It shows that the majority of women would breastfeed their children between one to two years.

Cut-off value for BIRTHIN

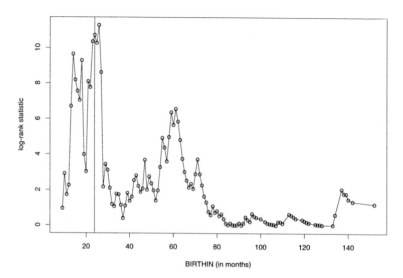

Figure 4.7: Log-rank statistic evaluated at all possible cut values for the preceding birth interval. The vertical line marks the cut-point at 24 months.

The two covariates measuring the preceding birth interval and the age of mother at birth are originally quasi-continuous and were binary coded at predefined cut-points. I deliberately abandoned optimised cut-points to avoid the known problem of inter-pretability and overestimation of their effects. However, I subsequently validated the chosen values by calculating the log-rank statistic for all possible cut-points, displayed in Figures 4.7 and 4.8. It shows, that a preceding birth interval of less that 24 months can definitely be considered as 'short', and also the categorisations of the age of the mother at birth at 21 and 35 are reasonable.

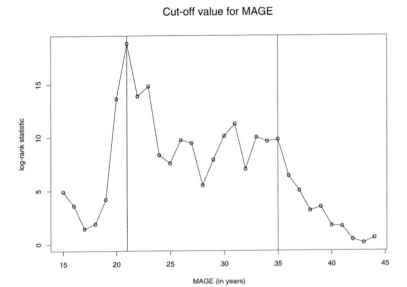

Figure 4.8: Log-rank statistic evaluated at all possible cut values for the mother's age at birth. The vertical lines mark the two cut-points at age 21 and 35.

4.6.3   Results of the survival analysis

Survival curves

To get a general idea of the data I started by analysing the survival estimates for different subgroups. Figure 4.9 shows the estimated survival rates depending on the child's and birth characteristics. It shows that first borns have a lower life expectancy than children not first born. Within the children not first born those who were born shortly after the previous delivery of their mother also show a low survival rate. At first sight, gender seems to have no impact on survival. If interaction with the order of birth is regarded the survival curves reveal, however, that within the first borns the girls clearly have a worse survival than boys.

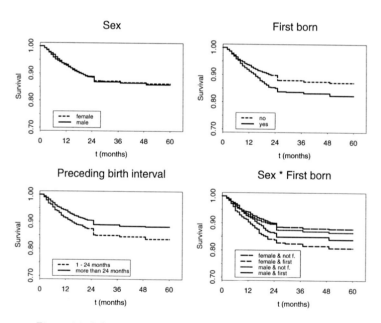

Figure 4.9: Infant mortality dependent on children's characteristics.

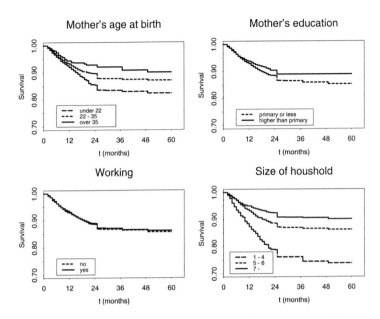

Figure 4.10: Infant mortality dependent on maternal characteristics and child's care.

The impact of the mother's characteristics are presented in Figure 4.10. It shows that older mothers and mothers with higher education improve the life expectation of a child. In contrast, the fact that a mother is currently working shows no impact. Rather it is the size of the household that again influences the mortality risk.

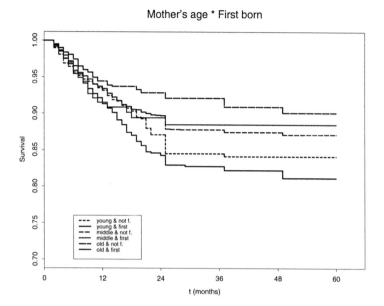

Figure 4.11: Infant mortality dependent on mother's age at birth and whether the child was first born.

In Figure 4.11 I additionally study the interaction between mother's age at birth and the first borns as there seems to be a natural association between these covariates. The survival curves reveal that it is more the young age of the mother than the fact that a child is first born which reduces survival chances. If the mother is between 21 and 35, their first borns have better survival chances. Hence, if the age of the mother would be included in a multivariate model, the effect of BFIRST is expected to decrease. Note that there is no woman in this data set, who was older than 35 at first delivery.

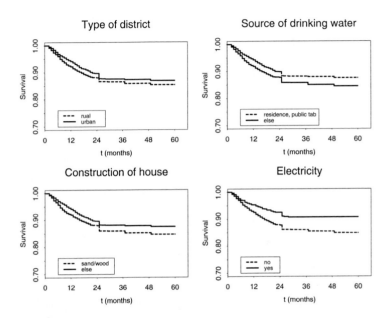

Figure 4.12: Infant mortality dependent on the constitution of the house.

Figure 4.12 shows the survival depending on type and condition of residence, where generally better housing standard positively influences survival of the child.

Univariate logit models

I then fitted univariate logit models with constant covariate effects to compare them with the time-varying estimates of the dynamic logit model. The posterior sample was generated by a block-wise Metropolis-Hastings algorithm with prior assumptions as described in Section 4.3 using the program BayesX (Lang & Brezger, 2001b). For tuning of the algorithm the average acceptance rate and the autocorrelation structure within the chain were evaluated in preliminary runs. Thereupon the distribution parameters of the inverse Gamma distribution $IG(a, b)$ were set to $a = 1$ and $b = 0.08$, expressing vague prior information of the RW2-variance. In addition the block sizes for sampling of the dynamic effects was chosen to lie between 3 to 12. After a burn-in phase of 2000 a posterior sample of 10000 replicates was generated. To avoid autocorrelation this sample was additionally thinned out by selecting only every tenth replicate. From the remaining sample characteristics of the posterior distribution were calculated as well as different model diagnosis criteria.

Table 4.6: Results for the constant effects from the univariate logit model

factor		β (post. mean)	post. StdDev	80 % CI
SEX		0.021	0.079	[-0.077 , 0.129]
BFIRST		0.326	0.091	[0.212 , 0.444]
BIRTHIN		-0.237	0.087	[-0.352 , -0.125]
MAGE	D1	-0.301	0.084	[-0.405 , -0.193]
MAGE	D2	-0.669	0.159	[-0.873 , -0.463]
BREASTF		-0.119	0.183	[-0.354 , 0.141]
MEDUCATION		-0.226	0.106	[-0.362 , -0.086]
MWORK		-0.0179	0.081	[-0.124 , 0.084]
HHSIZE	D1	-0.716	0.101	[-0.842 , -0.585]
HHSIZE	D2	-1.069	0.094	[-1.193 , -0.949]
URBAN		-0.167	0.079	[-0.271 , -0.064]
WATER		0.251	0.082	[0.1414 , 0.353]
HOUSE		-0.230	0.081	[-0.334 , -0.130]
ELECTRICITY		-0.490	0.119	[-0.640 , -0.347]

The results for the constant effects are summarised in Table 4.6, where the posterior mean of the parameters expresses the impact of the covariate. This is in agreement with the previously shown survival curves (Figures 4.9 - 4.12). Additionally the posterior standard deviation is given, together with the 10 % and 90 % credibility bounders enclosing 80 % of the posterior sample of β. For SEX, BREASTF and MWORK, the credibility bands include the zero, indicating low impact on survival if modelled with a constant effect.

Table 4.7: Criteria of the univariate models with constant effects

factor	$D(\overline{\mathcal{M}})$	$\overline{D}(\mathcal{M})$	DIC($\mathcal{M}$)	$df_{\mathcal{M}}$
NULL	7648.07	7675.99	7703.91	27.92
SEX	7651.17	7681.43	7711.68	30.25
BFIRST	7636.28	7666.13	7695.98	29.85
BIRTHIN*	5616.84	5640.23	5663.62	23.39
MAGE	7625.94	7656.06	7686.18	30.12
BREASTF*	7465.02	7495.00	7524.97	29.98
MEDUCATION*	7644.80	7673.83	7702.86	29.03
MWORK*	7651.76	7681.29	7710.83	29.53
HHSIZE	7529.37	7559.96	7590.55	30.59
URBAN	7645.16	7674.72	7704.28	29.56
WATER*	7614.12	7643.61	7673.10	29.49
HOUSE*	7615.61	7645.75	7675.88	30.14
ELECTRICITY*	7617.69	7646.91	7676.14	29.23

* = contains missing values

For those covariates that contain no missing values, the models can be validated by comparison of the DIC and the $\overline{D}$ with the null model. These criteria should decrease for models with covariates of a high explanatory value. In Table 4.7 the deviance criteria for the different models are given, i.e. the model deviance based on posterior estimates, the posterior mean of the deviance and the DIC. The models for SEX and MWORK, respectively, have a larger DIC than the null model, which only contains the baseline effect, which supports the finding that they carry no (constant) information. Also the explanatory value of type of district seems to be close to zero when modelled with a constant effect. In contrast, the order of birth, the age of the mother and the size of the household do influence survival of the child: these models have smaller values for all criteria. Note that the univariate model for the covariate

BIRTHIN is only presented to complete the picture. Since the birth interval is only given for children with older siblings, a comparison of this model to the null model does not make sense. For all other covariates, the deviance criteria do not allow a fair comparison with the null model, since these covariates contain missing values and hence the deviance, that is minus twice the sum of the log-likelihood over all available observations, will automatically yield a smaller value than the null model involving the full sample. As the main focus of this Section is on exploring dynamic effect structures, I omit a detailed analysis of the global effects of the covariates at this point of the analysis.

In addition to the deviance criteria the effective number of parameters are listed in the last column of Table 4.7. They represent the model dimension which is shown to be quite similar in all models and rises by about two in comparison to the null model, when a covariate is included with a constant effect. Note the exceptional value of $df_\mathcal{M}$ for the model including BIRTHIN, which might indicate a somewhat awkward dependency of estimate of model dimension on the number of observations.

Coming back to the major focus of this part of the analysis, that is the exploration of time-variation in the effect structures, regard the dynamic univariate fits, which are described in the following. The development of the effects over the age of the child resulting from univariate dynamic logit models are displayed in Figures 4.13 to 4.16. In addition the 80 % credibility region is given, as well as the constant estimates. For the covariates I only display the effect for the first 36 months since later the credibility intervals are too wide to provide any reliable information on the dynamic structure. The estimated baseline effect (in Figure 4.13) is similar in all univariate models and therefore only shown for the null model. The peaks jutting at month 24, 36 and 48 are caused by an outsized frequency of children reported with this age, surely a misrepresentation due to the way the data were collected.

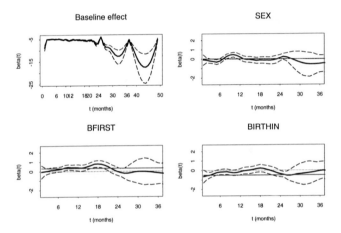

Figure 4.13: Estimated dynamic effects on infant mortality with 80 % credibility regions (- - - -). For comparison the constant effects (——) are given. The fine dashed line marks the zero.

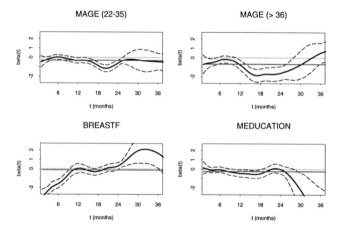

Figure 4.14: Estimated dynamic effects on infant mortality with 80 % credibility regions (- - - -). For comparison the constant effects (——) are given. The fine dashed line marks the zero.

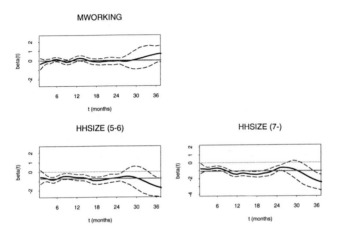

Figure 4.15: Estimated dynamic effects on infant mortality with 80 % credibility regions (- - - -). For comparison the constant effects (——) are given. The fine dashed line marks the zero.

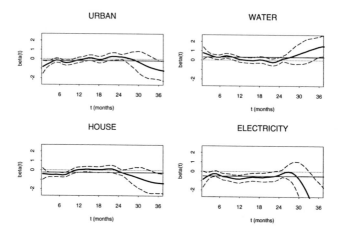

Figure 4.16: Estimated dynamic effects on infant mortality with 80 % credibility regions (- - - -). For comparison the constant effects (———) are given. The fine dashed line marks the zero.

Table 4.8: Criteria of the univariate models with dynamic effects

factor	$D(\overline{\mathcal{M}})$	$\overline{D}(\mathcal{M})$	$\mathrm{DIC}(\mathcal{M})$	$df_{\mathcal{M}}$
NULL	7648.07	7675.99	7703.91	27.92
SEX	7634.77	7673.15	7711.52	38.37
BFIRST	7625.43	7663.47	7701.51	38.04
BIRTHIN*	5612.80	5643.25	5673.71	30.46
MAGE	7596.25	7641.03	7685.81	44.78
BREASTF*	7432.09	7465.33	7498.58	33.24
MEDUCATION*	7640.09	7675.73	7711.38	35.65
MWORK*	7642.43	7679.96	7717.48	37.53
HHSIZE	7505.05	7550.11	7595.16	45.05
URBAN	7631.10	7668.74	7706.37	37.63
WATER*	7601.34	7638.92	7676.50	37.58
HOUSE*	7603.64	7640.83	7678.02	37.19
ELECTRICITY*	7605.78	7640.91	7676.05	35.14

* = contains missing values

Table 4.8 lists the values of the deviance criteria of the dynamic models. The model for MWORK, even though it contains missing values, yields larger values of $\overline{D}(\mathcal{M})$ and DIC than the null model, indicating that this covariate has no impact on survival. The missing impact of MWORK would support the hypothesis that children do not suffer nor benefit at any time, if the mother is working. Note, however, that MWORK indicates whether the mother is 'currently working' at the time of interview, which does not provide information of the past. In particular it does not necessarily apply for the time the child has lived. The covariate SEX, even when modelled with a dynamic effect, does not improve the fit sufficiently to reduce the DIC compared with the null model. The same is true for URBAN.

In Table 4.9 the constant and the dynamic fit are contrasted in order to decide if time-variation is substantial. The differences in the deviance criteria, $D(\overline{\mathcal{M}})$, $\overline{D}(\mathcal{M})$ and DIC are calculated by the criteria of the constant model minus the criteria of the dynamic model. In addition the Bayes factor $P(y|\mathcal{M}_{const})/P(y|\mathcal{M}_{dyn})$ and the posterior Bayes factor $\overline{L}(y|\theta, \mathcal{M}_{const})/\overline{L}(y|\theta, \mathcal{M}_{dyn})$ are evaluated. To facilitate the comparison of these criteria, they are given on the common $-2\log$-scale. The increase in the effective number of parameters is given by $-\Delta df_{\mathcal{M}}$. Hence in Table 4.9, a positive value denotes an improvement of the dynamic fit in comparison to the constant model, when evaluated by the corresponding criteria.

Table 4.9: Comparison of the univariate models with constant effects and with dynamic effects

factor	$-2\log(\text{PBF})$	$-2\log(\text{BF})$	$\Delta\overline{D}(\mathcal{M})$	$\Delta D(\overline{\mathcal{M}})$	$\Delta\text{DIC}(\mathcal{M})$	$-\Delta df_{\mathcal{M}}$
SEX	3.61	11.05	16.40	8.28	0.16	8.12
BFIRST	-4.80	6.56	10.86	2.66	-5.53	8.19
BIRTHIN	-11.31	-4.15	4.04	-3.02	-10.09	7.06
MAGE	21.28	22.32	29.69	15.03	0.37	14.66
BREASTF	34.20	32.23	32.93	29.66	26.39	3.27
MEDUCATION	-20.05	5.22	4.71	-1.91	-8.53	6.62
MWORK	-5.29	3.81	9.33	1.34	-6.66	7.99
HHSIZE	7.04	14.28	24.32	9.85	-4.61	14.47
URBAN	-1.95	11.75	14.05	5.98	-2.09	8.07
WATER	0.34	5.52	12.78	4.69	-3.40	8.09
HOUSE	-0.78	7.02	11.97	4.92	-2.17	7.05
ELECTRICITY	11.30	6.43	11.91	6.00	0.09	5.91

Unanimously the most distinct time-variation is seen for breastfeeding, causing a high risk of mortality when missing in the first year but having no impact when stopped within the second year (see Figure 4.14). This goes along with the general breastfeeding conduct of mothers in Zambia, which was illustrated in Figure 4.6. As described before, break-off by death or censoring of the child does not influence this result. Comparing the DIC of the constant model with the dynamic model shows the strong improvement of the fit.

Additionally the age of the mother at birth shows time-variation, which should be included into the model, indicating that the benefits of an older mother are rather relevant in the second year, probably after breastfeeding was stopped. The covariate ELECTRICITY, mirroring the condition of housing, influences survival within the very first months as well as later. The constant and the dynamic model fit equally well however, when comparing the DIC taking the model complexity into account. The dynamic effect of HHSIZE shows an increasing positive impact for rather large households in the period after the first year. This might express the particular importance of wealth and additional care in the period after breastfeeding. Note that this variation is not supported by the DIC. For BIRTHIN and MEDUCAT, in contrast to expectation, all criteria (but the non-penalised $D(\overline{\mathcal{M}})$) support the constant model, rejecting dynamic effects. The dynamic structures of the other factors are rather weak and, as indicated by the DIC, their improvements of the model fit do not justify the increased complexity given by $-\Delta df_{\mathcal{M}}$.

Multivariate Dynamic Logit Models

Subsequently multivariate models were fit jointly evaluating all factors. To reduce complexity in the multivariate model and due to the interpretation difficulties mentioned above, I omitted the covariate MWORK. Gender was included together with an interaction term with birth order (SEX*BFIRST), which has value one for first born boys and zero else. Time-varying effects were only assessed for BREASTF and MAGE.

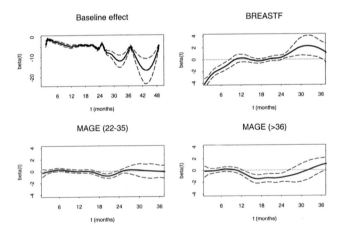

Figure 4.17: Dynamic effect for the multivariate model.

Figure 4.17 shows the baseline effect and the dynamic effects for breastfeeding and mother's age from the multivariate model. They are practically indistinguishable from the univariate results. This is different for the constant effects (Table 4.10). While the effect of the preceding birth interval, size of the household and education of the mother are basically the same, birth order looses its impact in the multivariate model and even changes signs. The covariates describing the condition of the residence, WATER, HOUSE and ELECTRICITY obviously carry partly identical information, since their effects are noticeably reduced. The 80 % credibility bands of the effects for HOUSE and WATER now include the zero. Also in the multivariate analysis, gender has generally no impact on mortality, expect for first borns where boys have a distinctly reduced risk.

Table 4.10: Results for the constant effects from the selected multivariate logit model

factor		post. mean	Post. StdDev	80 % CI
BFIRST		-0.028	0.149	[-0.216 , 0.161]
BIRTHIN		-0.290	0.099	[-0.419 , -0.165]
SEX		0.104	0.092	[-0.015 , 0.219]
SEX*BFIRST		-0.313	0.180	[-0.539 , -0.082]
DMEDUCATION		-0.167	0.122	[-0.325 , -0.014]
HHSIZE	D1	-0.662	0.104	[-0.799 , -0.535]
HHSIZE	D2	-0.963	0.096	[-1.086 , -0.842]
URBAN		0.151	0.126	[-0.007 , 0.319]
WATER		0.093	0.136	[-0.077 , 0.267]
HOUSE		0.001	0.125	[-0.158 , 0.158]
ELECTRICITY		-0.289	0.139	[-0.472 , -0.113]

To validate the dynamic effects I fitted additional models, where in turn the effects of BREASTF and MAGE were fixed constant, and compared these models using the six criteria as above. The values of the criteria are listed in Table 4.11 and again graphically illustrated in Figure 4.18. Clearly all criteria support the dynamic effects for these two factors, agreeing with the results of the univariate analysis. In addition I fitted the continuous variable for age of the mother at birth with an additive effect $\beta(AGE)$, i.e. an effect that is allowed to vary smoothly with the age of the mother. The criteria evaluating this model are presented in the last line of Table 4.11 and the last column in Figure 4.18, respectively. They show that this model can not compete

Table 4.11: Criteria for different multivariate models (w comprises all factors with fixed effects)

model	$-2\log(\text{BF})$	$-2\log(\text{PBF})$	$D(\overline{\mathcal{M}})$	$\overline{D}(\mathcal{M})$	$\text{DIC}(\mathcal{M})$	$df_{\mathcal{M}}$
$\mathcal{M}^*$	7320.26	7257.18	7220.65	7281.46	7342.27	60.81
$\mathcal{M}_1$	7349.52	7291.95	7255.26	7311.41	7367.56	56.15
$\mathcal{M}_2$	7337.07	7277.99	7249.74	7298.08	7346.43	48.34
$\mathcal{M}_3$	7520.12	7284.84	7256.83	7328.93	7401.04	72.11

$\mathcal{M}^* : \beta'w + \beta(t)BREASTF + \beta(t)MAGE$

$\mathcal{M}_1 : \beta'w + \beta\,BREASTF + \beta(t)MAGE$

$\mathcal{M}_2 : \beta'w + \beta(t)BREASTF + \beta\,MAGE$

$\mathcal{M}_3 : \beta'w + \beta(t)BREASTF + \beta(AGE)$

with the dynamic model $\mathcal{M}^*$. Still, it must be noted that by creating a categorical covariate subdividing age in three categories, model $\mathcal{M}^*$ already partially captures additive variations and additionally allows for time-varying effects. Based on the results of Table 4.11, the model allowing dynamic effects for BREASTF and MAGE is selected for further investigation.

Note that in Figure 4.18 the criteria show a parallel change over the models confirming the ordering that is expected by their formal definition. Only for the additive model, the DIC and $-2\log \text{BF}$ cross.

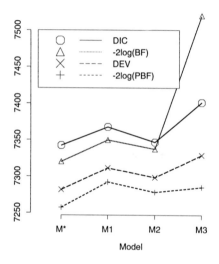

Figure 4.18: Comparison of the four multivariate models.

Reasons for stopping breastfeeding

Breastfeeding shows the strongest effect on survival, which induces a very high risk of mortality when stopped within the first month of life. One reason might be the low quality of substituting nutrition. However, we would then expect that the quality of water shows the converse time-structure, which it does not. Another explanation is instead given by an additional survey question in which the mothers reported on the reasons for which they stopped breastfeeding early. Possible reasons where e.g. illness or weakness of the mother, illness or weakness of the child, weaning age, pregnancy, working etc. .

Reasons for stopping breastfeeding

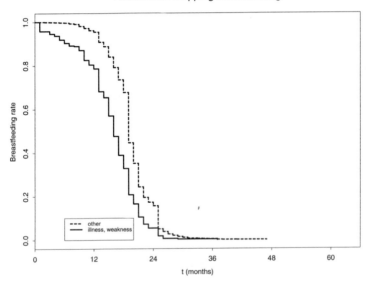

Figure 4.19: Duration of breastfeeding depending on reasons for stopping breast-feeding.

Figure 4.19 illustrates the dependence of duration of breastfeeding on the reasons for stopping, distinguishing illness and weakness of the mother or the child from innocuous reasons such as weaning age, pregnancy, working etc. . The analysis is thereto reduced to children which are already no longer breastfed and for which the stopping reason is known. The curves show that in Zambia early stopping of

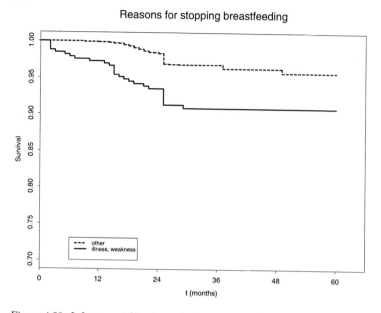

Figure 4.20: Infant mortality dependent on reasons for stopping breastfeeding.

breastfeeding was nearly always caused by illness or weakness, while healthy mothers with healthy children usually tend to breastfeed their children for quite a long time. Of the 2986 children of which we know that breastfeeding was stopped because of other reasons than illness or weakness, only eleven were breastfed for less than six months. The median duration of breastfeeding of these children is at 19 months and 75 % are breastfed over 17 months.

Figure 4.20 shows that when breastfeeding was stopped due to illness or weakness life expectancy of the child is clearly reduced compared to other stopping reasons. (Note that by explicitly having asked for the reason for which breastfeeding was stopped, the direction of the causality between the stopping reason and survival is clearly given, i.e. any illness of weakness of the child reported in this question was the cause of stopping breastfeeding and not vice versa.)

Consequently I included the stopping reasons in the selected model $\mathcal{M}^*$ from above. Similar to BREASTF I constructed a time-dependent covariate, STOPR, which is zero as long as the child is breastfed and changes to one in the month before breast-

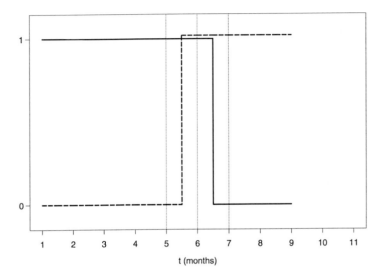

Figure 4.21: Construction of the time-dependent covariates BREASTF (———) and STOPR (- - - -).

feeding was stopped due to illness or weakness. For all other children it is zero throughout observation time. Note that STOPR is not identical to an interaction between a fixed covariate representing stopping reason and the time-dependent covariate BREASTF, since in the last moth of breastfeeding both covariates have value one when illness or weakness occurred.

Figure 4.21 illustrates the course of these two time-dependent covariates on the example of a child, who's mother stopped breastfeeding after six months due to illness. The covariate BREASTF has value one during the first six months and then zero, while the covariate STOPR is zero during the first five months and one from month six on. Hence, in month six, both covariates have value one.

In Figure 4.22 the dynamic effects of the covariates of this extended model are displayed. The effect of breastfeeding becomes much flatter revealing that the high risk of stopping breastfeeding within the first months can be explained by the stopping reasons. This is of course not visible when STOPR is included with a fixed effect

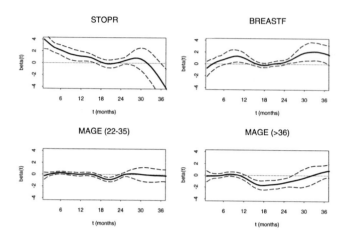

Figure 4.22: Dynamic effect for the multivariate model including STOPR.

(not shown). Table 4.12 lists the criteria for model comparison which are again graphically presented in Figure 4.23.

Table 4.12 lists the criteria for model comparison which are again graphically presented in Figure 4.23. All criteria advise to include STOPR with a dynamic effect. The high impact of STOPR, of course, is expected: if within the first months of life the child or the mother is so ill or weak that breastfeeding is stopped, it seems natural that due to this illness or weakness also mortality risk rises. By including STOPR into the model the dynamic structure of the effect of BREASTF could be corrected for this risk factor. However, STOPR obviously can not explain all time-variation of the effect of breastfeeding, since the criteria still reject a fixed effect for BREASTF.

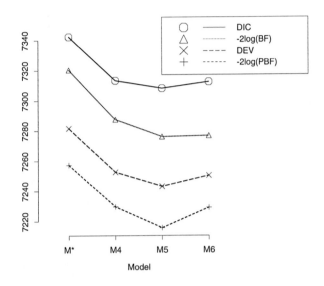

Figure 4.23: Comparison of the four multivariate models.

Table 4.12: Criteria for different multivariate models including STOPR ($\mathcal{M}^*$ denotes the selected model from Table 4.11)

model	$-2\log(\text{BF})$	$-2\log(\text{PBF})$	$D(\overline{\mathcal{M}})$	$\overline{D}(\mathcal{M})$	$\text{DIC}(\mathcal{M})$	$df_{\mathcal{M}}$
$\mathcal{M}^*$	7320.26	7257.18	7220.65	7281.46	7342.27	60.81
$\mathcal{M}_4$	7287.33	7229.47	7191.76	7252.48	7313.20	60.72
$\mathcal{M}_5$	7275.95	7215.62	7178.03	7243.06	7308.09	65.03
$\mathcal{M}_6$	7276.58	7229.03	7188.13	7250.25	7312.37	62.12

$\mathcal{M}^* : \beta'w + \beta(t)BREASTF + \beta(t)MAGE$

$\mathcal{M}_4 : \mathcal{M}^* + \beta\,STOPR$

$\mathcal{M}_5 : \mathcal{M}^* + \beta(t)STOPR$

$\mathcal{M}_6 : \beta'w + \beta(t)MAGE + \beta(t)STOPR + \beta\,BREASTF$

4.6.4   Summary

The study of infant mortality in Zambia provides a revealing example for a data situation where instead of a continuous time survival model a model for discrete time, such as the logit model, is most suitable. Moreover, the analysis demonstrates the importance of allowing for time-variation of the effects: The effect of breastfeeding is missed out if it is modelled in a time-constant fashion. However, the example also emphasises the need of deciding for each effect, whether it should be considered as fixed or as dynamic, in order to control model complexity. In the analysis twelve different covariates were studied and it was mainly for breastfeeding and age of the mother that a dynamic effect was needed, while for the other covariates, a constant effect was adequate.

For breastfeeding, the dynamic fit shows that it noticeably increases mortality risk of the child when it is stopped within the first year, while stopping later has no negative influence. The dynamic structure in the effect of breastfeeding, i.e. a strongly negative coefficient in the first months, can be explained by the reasons for which breastfeeding was stopped, where stopping due to illness or weakness of the mother or the child shows a high impact on survival. Note that the data provides no information on the general effect of short breastfeeding, since mothers in Zambia usually breastfeed their children for quite a long time. This was illustrated by the curves showing the duration of breastfeeding. Of the 2986 children of whom we know that breastfeeding was stopped due to other reasons than illness or weakness, only 11 were breastfed less than 6 months. Of course these 11 children do not allow for any inference.

The effect of the quality of the water does not, as one might expect, follow the converse time-structure of breast feeding. Possibly, it reflects sanitary conditions rather then influencing the quality of substituting food. Generally, it is seen that lower quality housing conditions increase infant mortality. In particular, electric connections in the residence have a positive influence on survival. It is only provided in a few houses indicating their high standard. A child also has better life expectancy when raised in a larger household. A large number of occupants might reflect greater wealth of the household as a rich household will attract occupants. Also, in a large household a child might benefit of the care of other adults. The impact of the size of the household should not be over-interpreted, however, since to some extent it also directly mirrors infant mortality: A household with high mortality will remain small. In addition, the size of the household refers to the actual size at the time of interview and not at birth of child. However, it can be assumed that the size of the household generally does not change that rapidly and that the categorical factor used in this analysis is rather stable. The impact of the age of the mother at birth supports that, at least in Zambia, an older mother can offer better care for the child.

Altogether this analysis reveals the different behaviour of the criteria when employed for model comparisons. Naturally, the non-penalised $D(\overline{\mathcal{M}})$ always prefers the more complex model. Also the posterior Bayes factor PBF weakly penalises model complexity; in nearly all univariate comparisons it favours the dynamic model. The Bayes factor, respectively its approximation by the posterior harmonic mean, is more selective. The DIC seems to penalise complexity most when comparing a constant and a dynamic fit, and in cases where the estimated number of parameters $df_{\mathcal{M}}$ was much larger for the dynamic fit, the approximation of the Bayes factor and the DIC even yield contradictory results. Possibly in our model context the approximation of the Bayes factor by the posterior harmonic mean is somewhat poor. When comparing models which distinctly differ, e.g. for the evaluation of the multivariate models, all criteria yield the same results.

Finally, it must be noted that the comparison of so many univariate and multivariate models for such a large data set was only possible because all criteria can be derived as a 'by-product' from the MCMC sampling of the posterior distribution of the parameters and do not demand much additional computation. Any sampling based methods for model diagnostics, e.g. the MCMC-estimation of the marginal likelihood proposed by Chib & Jeliazkov (2001) or cross-validation, are not feasible in situations of large data sets and complex model structures.

4.7 Discussion

In this chapter Bayesian approaches for the exploration of dynamic effect structures in survival data have been investigated. Survival data was discrete and treated as a binary time series. Dynamic logit models with semi-parametric predictors were employed, incorporating time-varying effects as well as time-constant effects. Smooth dynamic effect structures were modelled by a second order random walk. Inference on the model parameters resulted from Markov Chain Monte Carlo sampling, where in a hybrid algorithm for all model parameters an empirical posterior distribution is generated. This sampling algorithm only involves standard distributions, such as normal distributions and Inverse Gamma distributions. Beside the fixed covariate effects and the sequences of time-varying effects also the hyperparameters of the model are subject to posterior sampling. These include in particular the variance parameter of the second order random walk, which directly influences the smoothness of the dynamic effect structures. Thus, in this approach the degree of smoothness of the effect functions is determined by the data at hand.

A major focus of the chapter was Bayesian model selection whith a particular emphasis on the assessment if the fitted dynamic effect is actually required or not. Thereto different model criteria have been investigated, which allow for comparison of non-nested, arbitrarily complex models. These included the classical Bayes factor, the posterior Bayes factor, the posterior Deviance and the Deviance Information Criterion.

Since dynamic survival models are rather complex, computational aspects are of particular interest. A review of different proposals for the calculation of the Bayes factor showed that within the context of non-parametric survival modelling, most of them are unemployable. Only the approximation by the posterior harmonic mean was feasible. It was used in all following applications studied in this chapter. Like the posterior harmonic mean, also the posterior Bayes factor, the posterior Deviance and the DIC can be calculated directly from the MCMC output.

In the simulation study comparison of the different criteria revealed that their decisions between dynamic and constant effects do not fundamentally differ. Their behaviour in the different simulation settings were satisfactory and besides the Bayes factor, their sensitivity to prior assumptions is relatively small. However, the values of the criteria are somewhat shifted, where the DIC yields larger values than the posterior Bayes factor and naturally also larger values than the posterior mean of the deviance. The DIC also seems to penalise complexity more than the Bayes factor. This is in contrast to what is expected from the theoretical large sample approximation which will be considered in Chapter 5.

The infant mortality study has demonstrated in a nice way, how flexible and generally applicable these criteria are, allowing for the comparison of arbitrary complex, non-nested models.

In the next chapter I will investigate some major topics of the ongoing discussions of Bayesian model comparison in literature and discuss their concepts in more detail. In particular I will outline the relations, which can be drawn between the criteria.

Chapter 5

Discussion on Bayesian Model Diagnosis

In the following chapter I will illuminate the different Bayesian model criteria once again, 'critically but fair', reviewing the critics of different authors on these concepts. The different model criteria are then compared – based on their definitions and descriptions given in Chapter 4 – and their common base is explored.

5.1 Comments on the Bayes Factor

For many Bayesians the Bayes factor

$$BF_{0,1} = \frac{P(y|\mathcal{M}_0)}{P(y|\mathcal{M}_1)}$$

is the only adequate measure for model evaluation. Regarding the marginal likelihood as the predictive probability of observing the actual data, the Bayes factor can be viewed as measuring the relative success of model $\mathcal{M}_1$ and $\mathcal{M}_0$ at predicting the data.

Originally the Bayes factor was introduced by Jeffreys for the comparison of two competing scientific hypotheses. It is closely related to the likelihood ratio testing. In fact, in the simplest case, when the two hypotheses under consideration consist of single distributions with no free parameters ('simple versus simple') and no prior informations are involved, the Bayes Factor is equal to the likelihood ratio. In the case of unknown parameters θ the Bayes factor still has the form of a likelihood ratio in which the parameters are eliminated by integration rather than by maximisation as in Fisherian significance testing. Due to the origin of the Bayes factor theory in Bayesian hypothesis testing, the terms 'hypothesis' and 'model' are often used substitutionally.

The Bayes factor is also interpreted as a 'weighted likelihood ratio' of model $\mathcal{M}_1$ to $\mathcal{M}_0$, with the parameter priors being the weight functions, (Berger, 1985). The main difference to Fisherian hypothesis testing is that a p-value can only provide evidence against the defined null hypothesis, while the Bayes factor gives evidence 'in favour' of a hypothesis, either the null hypothesis or its alternative. This means that the Bayes factor is symmetric in its decisions. An additional advantage is that the definition of the Bayes factor is very general and does not demand alternative models to be nested, nor does it require any standard distributions. In particular Bayesian model selection based on the Bayes factor is consistent, i.e. if the true model is within the set of models under investigation, it will be selected (Berger & Pericchi, 2001). Thus, the Bayes factor may also be used to determine the posterior model probability that one model is correct (Kass & Raftery, 1995). Moreover, it can be employed as a weight for a composite estimate or model averaging, since it expresses the uncertainty of the model selection.

The use of the Bayes factor for model selection, however, has also serious drawbacks. A major limitation is its computational difficulties, which are particularly severe in complex models and large data sets. They have already been discussed in Chapter 4. Another critical property of the Bayes factor is its mentioned sensitivity to prior specifications. In some situations, this prior dependence might be of advantage, e.g. when prior distributions $\pi(\theta|M)$ are used to incorporate prior or external information into the model building process. In practice, however, such information is often not available. In such cases proper prior densities are hard to specify and the sensitivity of the Bayes factor to the choices of priors on the model parameters is then considered as a disadvantage. Thus, in such situations in Bayesin model estimation priors are often picked for convenience, knowing that if the sample is fairly large, then the impact of the prior is small. In testing based on the marginal likelihood this assumption can not be made.

A serious problem arises when improper (non-informative) priors are used and the parameter spaces of the models have different dimensions. Many authors argue, that improper priors can never be used with the Bayes factor, since under improper priors the marginal distribution is not defined. Jeffreys (1961) showed that under certain conditions the prior on the nuisance parameter is, however, much less relevant than the prior on the parameter at focus. He uses improper priors on nuisance parameters appearing in both null and alternative models. While this leads to an improper marginal distribution, the value of $P(y|\mathcal{M})$ for the given data remains well defined, so that this impropriety of priors does not disturb. Improper priors on parameters of interest are however problematic because, when used under $\mathcal{M}_1$ and not under $\mathcal{M}_0$, formally they force BF_{01} to become zero.

Even more critical is the use of proper but vague (diffuse) priors as they are cause of the Lindley paradox. Jeffreys (1961) already studied this problem and pointed out that diffuse (vague) prior distributions are not appropriate for hypothesis testing. Berger & Pericchi (2001) advise, "never use vague proper priors for model selection, but improper non-informative priors may give reasonable results". They note, that this also applies for vague proper priors on the hyperparameters in hierarchical models.

5.1.1 The Lindley Paradox

"A good paradox is one that is not easily resolved, ... In this respect, the Lindley paradox has been a particularly good one."

Stone (1997, p.263)

The Bayes factor is known for suffering from the Lindley paradox, which basically says that a point null hypothesis may strongly be rejected by a Fisherian significance test and yet be favoured by the Bayes factor based on small prior probabilities of the null hypothesis and a diffuse (vague) distribution over the alternative. This disagreement between Fisherian sampling-theory and Bayes theory was first studied by Jeffreys (1939/1961) and has been objective for repeated investigations and criticism, (e.g. Jeffreys, 1961; Lindley, 1957; Shafer, 1982; Aitkin, 1991). An example for the Lindely paradox is for instance found in Shafer (1982): Let y be a random variable that has a normal distribution with unknown mean μ and known variance σ^2, i.e. $y \sim N(\mu, \sigma^2)$. Consider a hypothesis problem where a point null hypothesis $H_0 : \mu = \mu_0$ is opposed to a composite alternative $H_1 : \mu \neq \mu_0$, that is

$$H_0 \ : \ \mu = \mu_0$$
$$H_1 \ : \ \mu \sim \pi(\mu)$$

and assume a vague prior distribution under the alternative, i.e. $\pi(\mu)$ is a fairly flat density with variance ϕ^2, and $\phi^2 \gg \sigma^2$. Consider now an observation y, where for simplicity y shall be univariate and might either be a single data point or an estimate of μ (e.g. $y = 1/N \sum_n y_n$). Suppose that the observation y has a value several σ away from μ_0. By Fisherian test theory, a value far from μ_0 is very unlikely under the null hypothesis and hence, in significance testing with the test statistic $(y - \mu_0)/\sigma$ such an observation would lead to the rejection of H_0. The Bayes factor regards the marginal likelihood of the two hypotheses, which in this case is for the point null hypothesis

$$P(y|\mathcal{M}_0) = \frac{1}{\sigma\sqrt{2\pi}} \exp\left(-\frac{(y - \mu_0)^2}{2\sigma^2}\right).$$

For the composite alternative hypothesis, which incorporates the prior distribution $\pi(\mu)$, the marginal likelihood is given by the integral

$$P(y|\mathcal{M}_1) = \int_{-\infty}^{\infty} \frac{1}{\sigma\sqrt{2\pi}} \exp\left(-\frac{(y - \mu)^2}{2\sigma^2}\right) \cdot \pi(\mu)d\mu.$$

The Bayes factor is the ratio of these marginal likelihoods. When ϕ^2/σ^2 is sufficiently large, $P(y|\mathcal{M}_1)$ will become relatively small and $BF_{0,1} > 1$. Thus the Bayes factor would not reject H_0 although the observation seems quit unlikely under the null hypothesis. The Lindely paradox can arise whenever an estimate of the parameter is very precise relative to the prior distribution $\pi(\theta)$, or in other words, if the variance of the estimate is much smaller than the variance of the specified prior distribution. Note that the precision of the estimate might thereby result from precise data as well as from a large sample.

The disagreement of the Fisherian significance testing and the Bayes factor results from the fact that the data supports rather strongly a relatively small interval of parameter values that does not include the null hypothesis. This information coming from the data is the only evidence taken into account in significance testing. Bayesian analysis additionally incorporates prior information. If the prior distributions over the alternative hypothesis are diffuse (vague) only a small probability is given to this interval supported by the data and the Bayes factor $P(y|\mathcal{M}_0)/P(y|\mathcal{M}_1)$ will be large. This gives the impression that the Bayes factor favours the (unrealistic) null hypothesis. However it seems to be more accurate to say, that in this situation the Bayes factor assigns very little evidence to the vague alternative prior. In other words, the Bayes factor rather rejects H_1 due to its imprecise prior specification then showing evidence *for* the null hypothesis H_0.

The actual background leading to a paradox is that the Bayesian (test) theory does not allow the interpretation of diffuse (vague) priors as representing lack of prior knowledge. This was clearly stated by DeGroot (1982), who wrote: "In summary, when diffuse prior distributions are used in Bayesian inference they must be used with care. Although they can serve as convenient and useful approximations in some estimations problems, they are never appropriate for tests of significance. Under no circumstances should they be regarded as representing ignorance."

Note that the definition of the dynamic survival model involves non-informative (improper) prior assumptions for the fixed covariate effects as well as for the initial values of dynamic effects. Formally, these parameters can't be seen as nuisance parameters as the models usually include different numbers of fixed and dynamic covariate effects. Still, in the simulation study as well as in the analysis of the infant mortality data, the Bayes factor behaves inconspicuously. This might be explained by the use of the posterior harmonic mean as an approximation of the Bayes factor. Also the Schwarz criterion, which approximates the Bayes factor, naturally does not suffer from the Lindely paradox.

5.1.2 Robust Bayes Factors

Many variants of the Bayes factor have been proposed to allow for a model evaluation based on the Bayes factor in the presence of uncertainty about the priors. These include the 'local Bayes factor' (Smith & Spiegelhalter, 1980) which gives increasing weights on a local neighbourhood of the null hypothesis. O'Hagan (1995) proposed the 'fractional Bayes factor', where a fraction ρ of the data is used to convert the prior to a proper posterior. The remaining $1 - \rho$ part of the data is then used to define the likelihood. The data for the prior-posterior transformation and the likelihood determination are thereby assumed to have the same distribution. The fractional Bayes factor is then determined using the fractional model likelihood

$$P_\rho(y|\mathcal{M}) = \int l^{(1-\rho)}(y|\theta, \mathcal{M}) \cdot P^{(\rho)}(\theta|y, \mathcal{M})d\theta.$$

This approach can be considered as an empirical Bayesian method, where a 'training sample' is used to form a proper prior which is given by the posterior of this training sample. This is then used in place of the vague prior to integrate the marginal likelihood.

Partitioning of the data to stabilise the Bayes factor has been proposed by other authors, too (see e.g. Discussion in Aitkin 1991). A difficulty with this method is, however, that the partitioning of the data influences the outcome of the criteria, as shown by O'Hagan (1991). Moreover, the partitioning of the data is clearly impractical if the data set is rather small. Spiegelhalter & Smith (1982) introduced a version of the Bayes factor based on an 'imaginary' training sample. The 'Intrinsic Bayes factor' by Berger & Pericchi (1996) uses a minimal training sample to convert improper priors to proper posteriors which are then used to define the Bayes factor for the remaining data. For stabilisation this is averaged over all possible minimal training samples. In complex modelling situations the definition of the minimal training sample is, however, not straightforward.

A related proposal is the 'pseudo Bayes factor' (Gelfand et al., 1992) which results from estimating the marginal likelihood (prior predictive density) $P(y|\mathcal{M})$ by the product of cross-validation predictive densities over all observations, i.e.

$$\prod_{n=1}^{N} P(y_n|y_{-n}, \mathcal{M}) ,$$

where y_n is the n-th observation and y_{-n} denotes all observations except y_n. The quasi Bayes factor is straightforward, however, it is computationally extensive.

Finally it should be noted, that minor changes of the Bayes factor due to different prior specifications are not of great importance as long as the prior sensitivity does not lead to contradictory conclusions. As Jeffreys noted: "We do not need Bayes factors with much accuracy. Its importance is that if BF > 1 (BF = $P(y|\mathcal{M}_0)/P(y|\mathcal{M}_1)$), the null hypothesis is supported by the evidence, if BF is much less than 1 the null hypothesis may be rejected. But BF is not a physical magnitude. Its function is to grade the decisiveness of the evidence." (Jeffreys, 1961, Appendix B).

5.2 Comments on the Posterior Bayes Factor

The posterior Bayes factor is defined as the ratio of the posterior expected likelihoods

$$\text{PBF}_{1,0} = \frac{\bar{l}(y|\mathcal{M}_1)}{\bar{l}(y|\mathcal{M}_0)},$$

where the posterior expectation of the likelihood $\bar{l}(y|\mathcal{M}) = \int l(y|\theta, \mathcal{M}) \cdot P(\theta|y, \mathcal{M}) d\theta$ is also regarded as posterior predictive distribution.

5.2.1 The Posterior Bayes Factor in Information Representation

Aitkin (1991) shows that $\bar{l}(y|\mathcal{M})$ can be viewed as an approximate penalised maximum likelihood criteria. Consider thereto the large sample situation of an approximate normal likelihood and negligible prior information. Let θ be a df-dimensional parameter and denote $L_\theta = \log l(y|\theta)$, omitting the model index $\mathcal{M}$. The first derivative L' gives the score function and the second derivative L'' is minus the observed Fisher information. Consider the expansion of the log-likelihood L about the maximum likelihood estimate $\hat{\theta}$

$$
\begin{aligned}
L_\theta &= L_{\hat{\theta}} + (\theta - \hat{\theta})' \left.\frac{\partial L}{\partial \theta}\right|_{\hat{\theta}} + \tfrac{1}{2}(\theta - \hat{\theta})' \left.\frac{\partial^2 L}{\partial \theta \partial \theta'}\right|_{\hat{\theta}} (\theta - \hat{\theta}) \\
&= L_{\hat{\theta}} + (\theta - \hat{\theta})' L'_{\hat{\theta}} + \tfrac{1}{2}(\theta - \hat{\theta})' L''_{\hat{\theta}} (\theta - \hat{\theta}).
\end{aligned}
$$

where for the maximum likelihood estimate $L'_{\hat{\theta}} = 0$, so that the likelihood can be approximated by

$$l(y|\theta) \approx l(y|\hat{\theta}) \cdot \exp\left\{-\tfrac{1}{2}(\theta - \hat{\theta})' {-L''_{\hat{\theta}}} (\theta - \hat{\theta})\right\}.$$

Taking the expectation with respect to the posterior distribution yields in

$$
\begin{aligned}
E_{\theta|y}[l(y|\theta)] &= \bar{l}(y) \\
&\approx l(y|\hat{\theta}) \cdot E_{\theta|y}\left[\exp\left(-\tfrac{1}{2}(\theta - \hat{\theta})' {-L''_{\hat{\theta}}} (\theta - \hat{\theta})\right)\right]. \qquad (5.1)
\end{aligned}
$$

With the Bayesian central limit theorem (4.5) the posterior distribution of the parameter θ asymptotically follows a df-dimensional normal distribution $\theta \sim N(\hat{\theta}, [-L''_{\hat{\theta}}]^{-1})$. Accordingly, for the large sample situation the posterior expectation in the second term of (5.1) can directly be calculated by

$$E_{\theta|y}\left[\exp\left(-\tfrac{1}{2}(\theta - \hat{\theta})' {-L''_{\hat{\theta}}} (\theta - \hat{\theta})\right)\right]$$

$$= \int \exp\left(-\tfrac{1}{2}(\theta - \hat{\theta})' - L''_{\hat{\theta}}(\theta - \hat{\theta})\right) \cdot (2\pi)^{-\frac{df}{2}} \left|-L''_{\hat{\theta}}\right|^{\frac{1}{2}} \exp\left(-\tfrac{1}{2}(\theta - \hat{\theta})' - L''_{\hat{\theta}}(\theta - \hat{\theta})\right) d\theta$$

$$= (2\pi)^{-\frac{df}{2}} \left|-L''_{\hat{\theta}}\right|^{\frac{1}{2}} \int \exp\left(-\tfrac{1}{2}(\theta - \hat{\theta})' - 2L''_{\hat{\theta}}(\theta - \hat{\theta})\right) d\theta$$

$$= (2\pi)^{-\frac{df}{2}} \left|-L''_{\hat{\theta}}\right|^{\frac{1}{2}} (2\pi)^{\frac{df}{2}} 2^{-\frac{df}{2}} \left|-L''_{\hat{\theta}}\right|^{-\frac{1}{2}}$$

$$= 2^{-\frac{df}{2}}$$

Hence for model $\mathcal{M}$ the posterior mean of the likelihood can be approximated by

$$\bar{l}(y|\mathcal{M}) = l(y|\hat{\theta}, \mathcal{M}) \cdot 2^{-df/2},$$

or on the familiar $-2\log$-scale it is

$$-2\log \bar{l}(y|\mathcal{M}) = D(\hat{\theta}, \mathcal{M}) + \log(2)\, df$$

with $D(\hat{\theta}, \mathcal{M}) = -2\log l(y|\hat{\theta}, \mathcal{M})$. And so $-2\log$ PBF is approximately equivalent to the maximum likelihood deviance penalised by $\log(2)$-times the dimension. As $\log(2) = 0.69$, the posterior Bayes factor penalises model complexity very little and will therefore tend to favour more complex models. Aitkin argues, that this penalty term arises naturally from the definition of the posterior Bayes factor and cannot be interpreted as an a priori penalty of complexity.

5.2.2 Criticisms of the Posterior Bayes Factor

The posterior Bayes factor has been subject to strong criticism. One issue is that it is using the data twice, once to obtain the posterior distribution of the parameters $P(\theta|y, \mathcal{M})$ and then again to rate the model by its posterior expected likelihood (see Discussion to Aitkin, 1991). Akaike (1991) writes thereto: "The repeated use of one and the same sample in the posterior mean of the likelihood function for the definition and evaluation of a posterior density certainly introduces a particular type of bias that invalidates the use of the mean as the likelihood of the model." Many discussants propose instead to split the data into two samples and estimate the posterior of the parameters from one part and the posterior of the likelihood form the other part, like e.g. the fractional Bayes factor described above.

Dempster (1997b) states that the major problem with the posterior Bayes factor is the particular summary statistic chosen to summarise the posterior probability, i.e. the mean of the posterior distributions of the likelihood. He proposes instead to regard the posterior mean of the logarithm of the likelihood ratio, that is $\log LR = \log(l(y|\theta,\mathcal{M}_0)/l(y|\theta,\mathcal{M}_1))$.

5.3 Comments on the DIC

The Deviance Information criterion is given by

$$\begin{aligned} \mathrm{DIC}(\mathcal{M}) &= \overline{D}(\mathcal{M}) + df_{\mathcal{M}} \\ &= D(\bar{\theta}|\mathcal{M}) + 2df_{\mathcal{M}}\,, \end{aligned}$$

with the effective number of parameters being $df_{\mathcal{M}} = \overline{D}(\mathcal{M}) - D(\bar{\theta}|\mathcal{M})$. The DIC particularly targets to cope with situations where the effective dimension of the model is not clearly defined, such as in hierarchical or non-parametric models. It follows the concept of other information criteria penalising the measure for goodness of fit by the model complexity. For large samples the DIC converges to the Akaike Information criterion $\mathrm{AIC} = D(\hat{\theta}) + 2df$.

Due to this relation to the AIC, the DIC cannot be asymptotically optimal in that it will not consistently select the 'true' model with increasing sample size, (Schwarz, 1978). Spiegelhalter et al. avert this criticism by arguing that they do not believe in a single true model.

In the proof of the approximation of the DIC by AIC, it was shown that

$$D(\theta) - D(\hat{\theta}) \approx -(\theta - \hat{\theta})' L''_{\hat{\theta}}(\theta - \hat{\theta})\,,$$

which is approximately χ^2_{df}-distributed (see Section 4.4, (4.39)). Thus, the posterior variance of $D(\theta) - D(\hat{\theta})$ is

$$V_{\theta|y}[D(\theta) - D(\hat{\theta})] = V_{\theta|y}[D(\theta)] \approx 2df\,. \tag{5.2}$$

Spiegelhalter et al. (2001) therefore mention, that the classical maximum likelihood deviance may be estimated by

$$\widehat{D}(\hat{\theta}) = E_{\theta|y}[D(\theta)] - \tfrac{1}{2}V_{\theta|y}[D(\theta)]\,,$$

calculating the empirical posterior mean and variance from a sample generated via MCMC.

Relation (5.2) however also suggests to use the posterior variance of the deviance as an estimate for twice the number of the parameters. This would give us the deviance information criterion as

$$\text{DIC}(\mathcal{M}) \approx E_{\theta|y}[D(\theta)] + \tfrac{1}{2}V_{\theta|y}[D(\theta)].$$

Recalling, that Dempster (1997b) proposed to compare two models by plotting the empirical posterior densities of their deviance, the DIC can be seen as a quantification of Dempster's idea by the location parameter of this density (i.e. posterior expectation) and its scaling parameter (i.e. posterior variance).

The effective number of parameters

Together with the DIC Spiegelhalter et al. introduce a measure for the effective model complexity, i.e. the effective number of degrees of freedom $df_{\mathcal{M}}$. This measure not only depends on the model definition, but moreover it depends on the data as well as on the parameter at focus. The authors argue that the effective number of parameters $df_{\mathcal{M}}$ should reflect the actual complexity of the model and hence these dependencies are justified. In the dynamic modelling context described in Section 4.2 this argument seems very reasonable. There, the dynamic effect functions are modelled by a random walk, where the variance of the random walk acts as a smoothing parameter and is estimated simultaneously with the model parameters. This corresponds to a data driven choice of the degree of smoothness. Thus, the complexity of these models depends on the data at hand, and the same should apply for its meassure, as it does for $df_{\mathcal{M}}$ by Spiegelhalter et al. .

However, in a full Bayesian analysis, as described in Section 4.3, the random walk variance is described by its posterior distribution. Consequently, it would be consistent also to consider a posterior distribution of the effective number of parameters, instead of calculating a one-point measure.

Another unclear issue is the behaviour of the measure $df_{\mathcal{M}}$ when formally the complexity of the model increases, e.g. when additional covariates or effect structures are added to the predictor. The simulation study showed that the number of effective parameters $df_{\mathcal{M}}$ may still decrease.

5.4 Comparison of the Different Criteria

Formally, the considered model criteria can be classified in 'prior' criteria, that is the Bayes factor, and 'posterior' criteria, that is the posterior Bayes factor and the Deviance criteria, which are derived from the posterior distribution. Practically in this thesis all criteria are calculated from the posterior sample, as I've used the posterior harmonic mean as an approximation of the Bayes factor. A closer look at the definition of the posterior Bayes factor, the deviance and the approximation of the Bayes factor reveals that they are all constructed from different posterior means of the likelihood $l(y|\theta, \mathcal{M})$.

The posterior Bayes factor PBF is defined as the ratio of the posterior expectations of the likelihood, which is approximated by

$$\bar{l}(y|\mathcal{M}) = \frac{1}{G} \sum_{g=1}^{G} l(Y|\theta^{(g)}, \mathcal{M}),$$

that is the posterior arithmetic mean of the likelihood based on the posterior sample of the parameters $\theta^{(1)}, \ldots, \theta^{(G)}$.

The posterior expectation of the deviance $\overline{D}(\mathcal{M})$, which has been proposed by Dempster (1997b) as a measure for goodness of fit and which is used in the Deviance Information Criterion, can be derived as the logarithm of the posterior geometric mean of the likelihood, since with $D(\theta, \mathcal{M}) = \log l(y|\theta, \mathcal{M})$ it is

$$\begin{aligned} \overline{D}(\mathcal{M}) &= \frac{1}{G} \sum_{g=1}^{G} \log l(y|\theta^{(g)}, \mathcal{M}) \\ &= \log \left[\prod_{g=1}^{G} l(Y|\theta^{(g)}) \right]^{\frac{1}{G}}. \end{aligned}$$

Finally, the Bayes factor is defined as the ratio of the marginal likelihoods $P(y|\mathcal{M})$, which can be approximated by importance sampling from the posterior likelihood, (see Section 4.4.2). This results in the posterior harmonic mean, i.e.

$$P(y|\mathcal{M}) \approx \frac{G}{\sum_{g=1}^{G} \frac{1}{l(Y|\theta^{(g)}, \mathcal{M})}}.$$

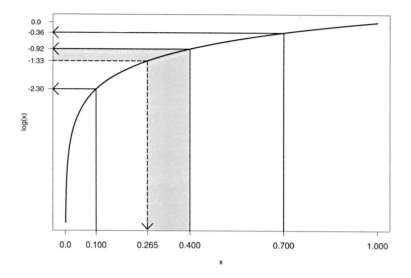

Figure 5.1: Comparison of the mean of x and the mean of log(x).

Thus, besides from their theoretical justifications, these criteria or their approximations, respectively, can also be seen as different posterior means of the likelihood.
Three consequences emerge from this representation:

1) In general the values of the arithmetic mean, the geometric mean and the harmonic mean for a particular sample $x_1, \ldots, x_N$ are not equal but follow the order

$$\bar{x}_{harm} \leq \bar{x}_{geo} \leq \bar{x}_{arith} . \tag{5.3}$$

This ordering of the values of the criteria is found in the evaluation of the multivariate models for the infant mortality data in Section 4.6.

2) The order (5.3) implies that it is not appropriate to use the same interpretation
schemes for the different criteria.

3) The different means do not yield in the same ranking when used to compare
two models. This inequality in ranking is illustrated in Figure 5.1, where two
observations, 0.1 and 0.7, are considered, resulting in a mean value of 0.4. The mean
value of their logarithms, i.e. the logarithm of the geometric mean, is $\frac{1}{2}\big(\log(0.1) + \log(0.7)\big) = -1.33$ which is smaller than the logarithm of their arithmetic mean, i.e.

$\log(0.4) = -0.92$. Now all pairs of values which result in a mean and a mean of the logarithm that lies within the shaded area show contradictory ranking. These are for example all pairs of values that lie themselves in this area. A similar graphical sketch could be drawn to compare the harmonic mean with the arithmetic mean or the geometric mean.

Finally, note that all criteria that are derived from the posterior sample are, of course, also subject to MCMC sampling errors.

5.5 Comparison by Large Sample Properties

In the large sample situation with negligible prior information and approximate normal likelihood, all four criteria – the Bayes factor, the posterior Bayes factor, the posterior mean of the deviance and the Deviance Information criterion – can be approximated by representations of maximum likelihood Information criteria, as it has been described in Section 4.4. Table 5.1 summarises these approximations, where $\hat{\theta}$ denotes again the maximum likelihood estimate and df is the real model dimension.

Table 5.1: Approximations by Information criteria

criterion		IC-approximation	
$\text{BF}(\mathcal{M})$	$\rightarrow$	$\text{BIC}(\mathcal{M}) =$	$D(\hat{\theta}) + \log(N^*)df$
$\text{PBF}(\mathcal{M})$	$\rightarrow$	$-2\log(\text{PBF}) \approx$	$D(\hat{\theta}) + \log(2)df$
$\overline{D}(\mathcal{M})$	$\rightarrow$	$\overline{D}(\mathcal{M}) \approx$	$D(\hat{\theta}) + df$ *
$\text{DIC}(\mathcal{M})$	$\rightarrow$	$\text{AIC}(\mathcal{M}) =$	$D(\hat{\theta}) + 2df$

* see Section 4.4, (4.40)

The Bayes factor can be approximated by the Schwarz criterion, from which the Bayesian Information criterion BIC is derived. It penalises the deviance evaluated at the maximum likelihood estimate by $\log(N^*)$-times the model dimension, where N^* is the number of independent observations. With BIC also the Bayes factor takes model dimension into account.

The DIC can be approximated by the Akaike Information criterion, which penalises the maximum likelihood deviance $D(\hat{\theta})$ by twice the number of model parameters.

Thus, the DIC penalises model complexity less than the BIC (given that the number of observations is larger than $\exp(2) = 7.4$) and therefore is theoretically expected to choose higher-dimensional models compared to the Bayes factor or the BIC, respectively. The larger the number of observations, the bigger the difference in the selection based on one or the other criteria. (Note that this order of penalisation is only true when the conditions for the approximation are fulfilled.)

The posterior mean of the deviance $\overline{D}$ can be approximated by $D(\hat{\theta})$ penalised by the model dimension, (see Section 4.4, (4.40)). Thus, it penalises complexity less than the BIC for $log(N) > 1$ and – per definition – less than the DIC($\mathcal{M}$).

Minus twice the posterior Bayes factor is for large samples approximately equivalent to the maximum likelihood deviance plus $\log(2)$-times the model dimension and penalises complexity the least.

Thus, for a fixed model complexity df (and more than seven observation) the BIC will yield the largest value, while the $\overline{D}(\mathcal{M})$ will be smaller and the PBF will have the smallest values. Recalling that the Bayes factor can also be approximated by the harmonic mean, while the posterior Bayes factor is defined as the arithmetic mean and $\overline{D}(\mathcal{M})$ is the logarithm of the geometric mean, this fits to the ordering given in (5.3).

I close the discussion of the criteria by noting that despite their different theoretical backgrounds and the different justifications of their concepts, they are related to each other on a common base.

Chapter 6

A Comparative Study: Prognostic Factors in Breast Cancer

In this chapter dynamic effect structures of prognostic factors are explored as part of a breast cancer study on prognosis of disease free survival. In addition covariate selection is performed leading to an optimal semi-dynamic predictor, which incorporates all important factors at hand and includes dynamic effects structures only when clear evidence for time-variation is shown. Rather than postulating a new prognostic system, the focus of this chapter will be on investigating the performances of the Fractional Polynomials approach and of the Bayesian approach in a comparative application. Thus, two analyses of the same data set are presented, one fitting a dynamic Cox model and the other fitting a dynamic logit model, using the methods described in the chapters above. This application shall reveal the similarities and the discrepancies of the proposed strategies.

For the Fractional Polynomial approach a backfitting-type algorithm was proposed in Section 3.2.3 to fit multivariate dynamic Cox models. This algorithm already includes a likelihood ratio test in each step that decides for each covariate if it requires a dynamic effect or not. Within the Bayesian framework, the effects of all covariates are estimated simultaneously using Markov Chain Monte Carlo methods. Here, the decision between a time-constant and a dynamic effect will have to be done manually by comparing models of different complexity using the model criteria which were proposed in Section 4.4. Practically this means that in a stepwise selection procedure successively a row of multivariate models are fitted, including fixed and time-varying effects.

Besides the handling of dynamic effects, also the behaviour of the approaches in covariate selection will be studied, where for each covariate its global impact is investigated after having decided upon its optimal effect structure. In the Fractional Polynomial approach this is realised by employing an adjusted likelihood ratio test while in the Bayesian approach again different models are compared using Bayesian model criteria. Finally, an additional covariate will be included into the model, which, from the medical view, is assumed to carry partially equivalent information as one of the other covariates. This will reveal how parsimonious the different approaches are.

The chapter starts by outlining the general background of the breast cancer study and introducing the data set in Section 6.1. In Section 6.2 the Fractional Polynomials approach is applied to analyse the data within the framework of dynamic Cox modelling. The Bayesian approach for discrete survival data is employed in Section 6.3, where different multivariate dynamic models are fitted. Finally in Section 6.4 the results of the two approaches are compared with respect to the selected predictor as well as to the fitted dynamic curves.

6.1 The Data

To date, breast cancer cannot be cured once distant metastases have occurred. Yet, even within the group of node negative breast cancer patients, i.e. patients whose axillary lymph nodes showed no tumour cell involvement at time of surgery, the 10 years relapse rate is still about 30 % (see Figure 6.1). This means that nearly every third node negative patient suffers from recurrence of the disease and eventually dies of metastasis. It is therefore of great interest to identify these high-risk patients as early as possible for further systematic adjuvant therapy. However, accurate identification of this risk group has not been possible using traditional histological and clinical factors alone. In the last decades research on novel tumour-biological factors became relevant which are directly measured in tumour tissue. It is hoped that a further knowledge of the functionality of these factors eventually provides a basis for a better understanding of the disease. Considerable contributions have already been made, many of them with a particular focus on the assessment of the prognostic impact of newly detected tumour-biological factors. Two of these tumour-biological factors, the urokinase-type Plasminogen Activator uPA and its type-1 inhibitor PAI-1, have been introduced in Section 3.4. Their relevance for

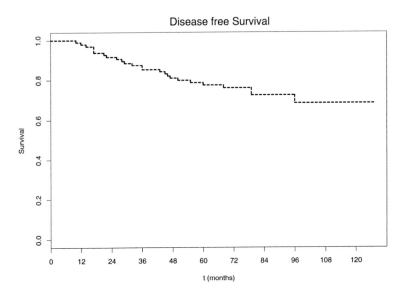

Figure 6.1: Disease free survival of node negative breast cancer patients.

breast cancer prognosis has already been verified in various long-term studies (e.g. Harbeck et al., 1999; Harbeck et al., 2000).

One of the currently investigated factors is the protein T1-S. It is observed to be higher expressed in tumour tissue than in normal tissue. Prechtel et al. (2001) have studied its association with disease free survival of node negative breast cancer patients in a group of 102 women who showed no distant metastases at time of surgery. Their results provide a first indication that a higher T1-level reduces the risk of relapse.

In the following analyses the same data set is used to explore dynamic structures in the effect of T1-S and other prognostic factors. The main focus is on providing a comparative study of the performance of the approaches for dynamic survival modelling and model selection that have been presented in this thesis. For this reason the analysis will be restricted to a small, pre-selected set of factors, which are known to be of prognostic interest from previous publications. Table 6.1 gives an overview of the factors and their interpretation. To study dynamic effect structures

Table 6.1: Prognostic factors of the breast cancer study

factor	range	coding	interpretation
TUMOUR	0.5 cm - 7 cm	0: $\leq$ 2 cm 1: $>$ 2 cm	tumour size
HORMONE		0: negative 1: positive	hormone receptor state
T1	0 - 12	0: $\leq$ 4 1: $>$ 4	T1-S score
MIB1	0 % - 90 %	0: $\leq$ 25 % 1: $>$ 25 %	proliferation associated antibody MIB1
GRADING	1 - 3	0: 1 or 2 1: 3 or 4	tumour grading

all factors are coded in a binary fashion.

One of the classical prognostic factors is tumour size. It is given in centimetre and dichotomised at the common cut-point of 2 cm. The hormone receptor state is determined from the estrogen and progesterone receptor expression and is positive if at least one of these receptors is positive. The new factor T1-S is given by a score ranging from 0 to 12. It specifies the level of the protein T1 in tumour tissue which is considered to be overly expressed if the score is larger than 4. The antibody MIB1 quantifies the antigen Ki-67, which is known to influence proliferation, that is the dispersion of the tumour within the surrounding tissue. MIB1 is also measured within the tumour tissue. It is given by the percentage of tumour cells expressing the antibody and a cut-point is set at 25 %. In a second step of the analysis tumour grading will be introduced in the analysis. It is given by a score ranging from 1 to 4, which is assigned by a pathologist, who classifies the tumour with respect to its proliferation characteristics. Hence, it is suspected to carry similar information on prognosis as MIB1.

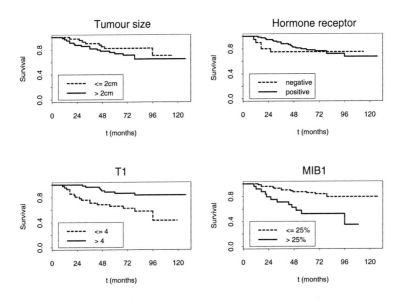

Figure 6.2: Disease free survival in different subgroups.

Complete data of these five covariates was available for 97 patients of whom 25 had a relapse and 20 died during follow-up. Median follow-up of the patients still alive at the end of observation period was 82 months. Figure 6.2 illustrates the disease free survival for different subgroups formed by the binary coded factors. The disease free survival rates for low and high T1 clearly differ. The same applies for MIB1, whereas the curves for the factor TUMOUR are rather close. The survival curves for positive and negative hormone receptor state cross, giving a first indication for underlying dynamic effect structures.

6.2   Fitting Fractional Polynomials

To fit a multivariate dynamic Cox model based on Fractional Polynomials the backfitting-type algorithm of Section 3.2.3 is used. It includes a data-driven decision on time-variation for each factor where the significance level was set to $\alpha_{PH} = 0.05$. The powers of the Fractional Polynomials are selected from the power set $P \in \{-2, -1, -0.5, 0, 0.5, 1, 2\}$.

Table 6.2: Results of the multivariate FP model for the breast cancer data

factor: $\beta_i(t)$	m	$H_0 : \beta_i(t) = \beta_i$ p-value	$H_0 : \beta_i(t) = 0$ p-value (adj.)
TUMOUR: $\beta = 0.6$	0	—	0.281
HORMONE: $\beta(t) = -4.3 + 0.2t$	1	0.005	0.028
T1: $\beta(t) = -31.5 + 9.6 \log(t) - 0.002t^2$	2	0.044	0.008
MIB1: $\beta = 1.0$	0	—	0.038

The resulting effect structures and the corresponding p-values of the test on time-variation, i.e. the test on the PH-assumption, are given in Table 6.2. In addition the adjusted p-values of the global test $H_0 : \beta(t) = 0$ are listed, which assesses the impact of the covariates on disease free survival.

The factor HORMONE shows a strong violation of the assumption of constant effects with $p_{PH} = 0.005$. The selected Fractional Polynomial is a strongly increasing, linear function over time. Its shape is pictured in Figure 6.3. In addition to the effect of HORMONE, also the effect of T1 shows a change over time. It is described by a Fractional Polynomial of degree two, which results in a bent function. Its negative effect in the beginning indicates that low values of T1-S increase the risk of relapse. This effect approaches zero over time. Towards the end of follow-up, the standard error bands of the effect of T1 become quite wide, as only little information remains at these time points. Although these error bands still include the zero, the effect $\beta(t)$ shows a tendency to decrease again. In contrast, MIB1 shows a constantly high impact on disease free survival with a relative risk for patients with high MIB1 of 2.77. The factor TUMOUR also shows no evidence for dynamic structures. However, the adjusted p-value of the test on the global effect of TUMOUR is 0.281, revealing that is does not contribute significant information to the model. This is in agreement with the plotted 2-standard error bands shown in Figure 6.3, which include the zero.

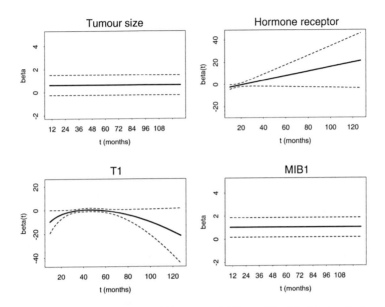

Figure 6.3: Estimated effects for the optimal FP model with ± 2 standard error bands (- - - -).

If tumour size is omitted from the model, the effects of the remaining three factors and their standard deviations remain unchanged, i.e. $\beta_{HORMONE}(t) = -4.3 + 0.2t$, $\beta_{T1}(t) = -31.1 + 9.8 \log(t) - 0.002t^2$, $\beta_{MIB1} = 1.0$.

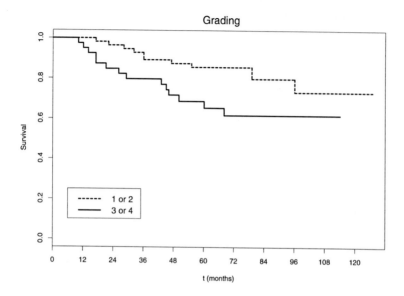

Figure 6.4: Disease free survival by tumour grading.

Table 6.3: Results of the multivariate FP model including GRADING

factor: $\beta_i(t)$	m	$H_0 : \beta_i(t) = \beta_i$ p-value	$H_0 : \beta_i(t) = 0$ p-value (adj.)
TUMOUR: $\beta = 0.6$	0	—	0.333
HORMONE: $\beta(t) = -4.0 + 0.2t$	1	0.006	0.033
T1: $\beta(t) = -31.1 + 9.5\log(t) - 0.002t^2$	2	0.05	0.007
MIB1: $\beta = 0.9$	0	—	0.094
GRADING: $\beta = 0.5$	0	—	0.538

To examine the behaviour of the Fractional Polynomial approach in variable selection, tumour grading is additionally included in the model. Figure 6.4 gives the
disease free survival in the two subgroups formed by the dichotomous factor GRAD-
ING. The two curves distinctly differ showing a higher relapse risk for patients with
tumour grading 3 or 4. Since grading is a staging of the tumour with respect to
its proliferation characteristics, it is of interest to examine, if it contributes any
additional information to prognosis besides MIB1. The results of the multivariate
FP model including GRADING are given in Table 6.3, where GRADING shows no
additional significant effect ($p_{total} = 0.538$). At the same time, the adjusted global
p-value of MIB1 rises to $p_{total} = 0.094$ indicating that both factors partially carry
equivalent information.

6.3   Fitting a Bayesian Logit Model

The breast cancer data are now analysed within the Bayesian framework using the
dynamic logit model with state space structure for discrete survival time data as
introduced in Chapter 4. The transition model is given by a second order random
walk where the variance parameter is a-priori assumed to follow the highly dispersed
inverse Gamma distribution $IG(0.001, 0.00001)$. Multivariate models are fitted by
the MCMC-algorithm described in Section 4.3. Block-sizes for the time-varying
effects range from 2 to 4, where in every MCMC-iteration the bounds are randomly
changed. After a burn-in period of 5000, a chain of 30000 elements is generated
yielding an average acceptance rate of 70 %. Autocorrelation within the final sample was reduced by selecting only every 10th element before performing posterior
estimation.

To decide within this framework which of the covariates require a dynamic effect
and which may be fixed, different multivariate models are fitted, which include time-
constant and time-varying effects. These are compared by employing the proposed
Bayesian model criteria of Section 4.4, that is the posterior mean of the deviance $\overline{D}$,
the Bayes factor BF, the posterior Bayes factor PBF and the Deviance Information
criterion DIC. The model deviance $D(\overline{\theta})$ is required for the calculation of the DIC
and therefore listed as well. However, as described before, it generally decreases
with increasing model complexity and is thus of no interest for a model selection
process on its own.

The upper part of Table 6.4 shows the values of the criteria for the static solutions,
that is the null model which includes no covariates beside the baseline effect, the constant model, assuming all covariate effects to be constant and the fully time-varying
model, allowing all covariate effects to change over time. All criteria coincide in
that the covariates provide substantial information on disease free survival. Moreover they confirm that allowing for dynamic effects structures improves the fit of
the model.

The last column of Table 6.4 lists the effective number of parameters as defined
in Section 4.4, i.e. $df_{\mathcal{M}} = D(\overline{\theta}) - \overline{D}$. For the null model, which only considers
the dynamic baseline effect, model complexity is 2.49. When the four constant
effects are included, the complexity increases by 4.02, which would correspond to
one additional parameter per constant effect. If all four effects are allowed to vary
over time, model complexity rises to 12.77.

Table 6.4: Bayesian model criteria for different multivariate models for the breast cancer study

factors:	TUMOUR	HORMONE	T1	MIB1	$D(\bar{\theta})$	$\bar{D}$	$-2\log(\mathrm{BF})$	$-2\log(\mathrm{PBF})$	DIC	$df_{\mathcal{M}}$
Global Solutions										
	—	—	—	—	255.49	257.98	260.62	257.18	260.47	2.49
	β	β	β	β	236.98	243.49	249.75	241.34	250.00	6.51
	$\beta(t)$	$\beta(t)$	$\beta(t)$	$\beta(t)$	219.28	232.05	243.48	226.60	244.82	12.77
$\mathcal{M}^*$	β	$\beta(t)$	$\beta(t)$	β	220.35	230.59	240.21	226.61	240.83	10.24
GRADING										
	β	$\beta(t)$	$\beta(t)$	β	219.17	230.21	239.24	225.88	241.25	11.04
	$\beta(t)$	$\beta(t)$	$\beta(t)$	β	219.09	231.34	243.61	226.41	243.59	12.25

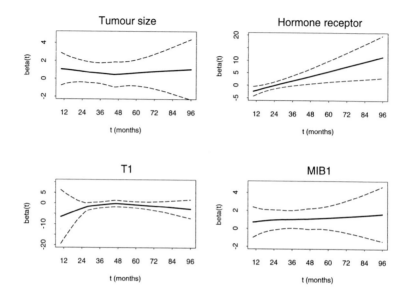

Figure 6.5: Estimated effects for the fully varying Bayesian model with ±2 posterior standard error bands (- - - -).

The four time-varying covariate effects of the fully varying model are displayed in Figure 6.5 offering a first insight into the dynamic effects structures. Instead of plotting credibility regions as in Section 4.6, the posterior mean $\bar{\beta}(t)$ is given together with ±2 posterior standard error bands. These allow for a better comparison with the results of the Fractional Polynomial approach. The graphs present a similar picture as found for the dynamic Cox model. The effect of the hormone receptor state strongly increases over time while the course of the effect of T1 is bent. The effects of tumour size and of MIB1 show no noticeable time-variation.

In Figure 6.6 the baseline effect for relapsing after operation is plotted. Generally the risk for relapse decreases over time. Note that the baseline effect was explicitly included in the dynamic logit model, while its computation is circumvented within the partial likelihood framework of the Cox model.

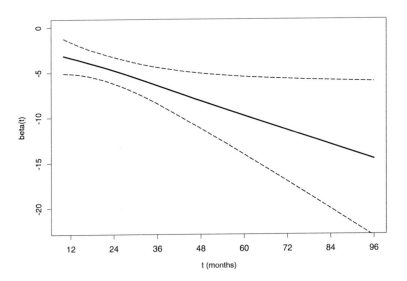

Figure 6.6: Estimated baseline effect for the optimal selected model $\mathcal{M}^*$ with ± 2 posterior standard error bands (- - - -).

Table 6.5: Bayesian model selection for the breast cancer study

factors:	TUMOUR	HORMONE	T1	MIB1	$D(\hat{\theta})$	$\bar{D}$	$-2\log(\text{BF})$	$-2\log(\text{PBF})$	DIC
Forward Selection									
	β	β	β	β	236.98	243.49	249.75	241.34	250.00
step 1	$\beta(t)$	β	β	β	236.12	243.93	252.12	241.34	251.74
	β	$\beta(t)$	β	β	227.54	235.00	**241.79**	**232.63**	**242.46**
	β	β	$\beta(t)$	β	230.79	239.73	249.88	235.72	248.67
	β	β	β	$\beta(t)$	236.78	244.17	255.45	241.71	251.56
step2	$\beta(t)$	$\beta(t)$	β	β	227.25	235.90	246.16	232.96	244.55
	β	$\beta(t)$	$\beta(t)$	β	**220.35**	**230.59**	**240.21**	**226.61**	**240.83** $\mathcal{M}^*$
	β	$\beta(t)$	β	$\beta(t)$	227.57	236.52	243.88	233.47	245.47
step 3	$\beta(t)$	$\beta(t)$	$\beta(t)$	β	**219.76**	**231.68**	**241.64**	**226.55**	**243.60**
	β	$\beta(t)$	$\beta(t)$	$\beta(t)$	221.80	233.44	244.55	228.78	245.08
step 4	$\beta(t)$	$\beta(t)$	$\beta(t)$	$\beta(t)$	**219.28**	**232.05**	243.48	**226.60**	244.82
Backward Deletion									
	—	$\beta(t)$	$\beta(t)$	$\beta(t)$	**222.21**	**232.29**	243.54	**228.16**	**242.37**
	β	—	$\beta(t)$	$\beta(t)$	230.48	238.91	252.99	234.94	247.34
	β	$\beta(t)$	—	β	234.98	241.51	248.16	239.32	248.04
	β	$\beta(t)$	$\beta(t)$	—	225.66	234.97	243.57	231.31	244.28

To take a data-driven decision, which of the effects distinctly vary over time, a series of models is fitted in a stepwise forward selection procedure. These are compared by employing the different Bayesian model criteria. The models which were fitted in this procedure and their values of the criteria are listed in Table 6.5. The procedure starts with the constant model. In the first step a dynamic effect is included for one covariate at a time. All criteria find the strongest improvement of fit when the effect of HORMONE is allowed to vary with time. In the second step an additional dynamic effect is included, where all criteria again unanimously decide on T1. When a third dynamic effect is considered in the next step, only the model deviance $D(\bar{\theta})$ further decreases. However, as mentioned above, it generally becomes smaller with increasing model complexity and thus has its minimum for the fully varying model. The posterior Bayes factor is basically unchanged when including an additional dynamic effect for TUMOUR or even for all four factors. As the general idea is to maintain parsimony, model $\mathcal{M}^*$ is selected. It includes dynamic effects for the factors HORMONE and T1 and minimises the Bayes factor, the posterior mean of the deviance $\overline{D}$ and the Deviance Information criterion DIC.

Next in a backward deletion step four more models are fitted by omitting one factor at a time from model $\mathcal{M}^*$. For all four factors this yields a reduction of the goodness of fit so that model $\mathcal{M}^*$ is retained.

In Table 6.5 the smallest values are highlighted by bold large letters for each criteria. Since for the posterior Bayes factor the three smallest values only differ in the second digit, no optimal model is highlighted. All models that yield values which differ less than 2 from $\mathcal{M}^*$ are printed in bold letters. These are the models, that show no 'strong' inferiority compared to the optimal model $\mathcal{M}^*$, when applying the interpretation schemes given in Section 4.4. Comparison of these equally valued models confirms that the Deviance Information criterion rather tends to select more parsimonious models while the other criteria also consider the model including a dynamic effect for tumour size or even the fully varying model on par with model $\mathcal{M}^*$. Note, that for tumour size only the Bayes factor provides 'strong evidence' that it improves the fit, while the other three criteria rate the model without the factor TUMOUR as equally good.

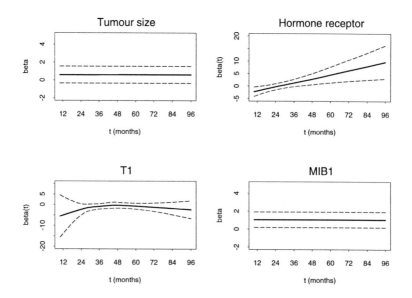

Figure 6.7: Estimated effects for the optimal model $\mathcal{M}^*$ with ± 2 posterior standard error bands (- - - -).

Figure 6.7 shows the effects of model $\mathcal{M}^*$. The dynamic structure of hormone receptor state and T1 did not change when fixing the other two effects. Note, however, that by restricting the effects of tumour size and MIB1 to time-constancy, their error standard bands become much narrower. Table 6.6 lists the effective number of parameters $df_\mathcal{M}$ for the models considered during the model selection process. It shows that including a dynamic effect for HORMONE requires about one additional effective parameter compared with the constant model. This matches with the linear structure of its time-dependency, which was plotted in Figure 6.7. When in addition the effect of T1 is allowed to vary with time, the effective number of parameters increases by 2.78. Obviously the bent dynamic effect structure required for T1 increases complexity to a greater extent.

In the last column of Table 6.6 0.5-times the posterior variance of the deviance is given. As pointed out in Section 4.4.5, under normality of the likelihood and negligible priors this approximately corresponds to the effective number of parameters. In our example the values are in general quite close to the values of $df_\mathcal{M}$. The difference

Table 6.6: Effective number of model parameters for the Bayesian dynamic logit models for the breast cancer study

factors:		HORMONE	MIB1	$df_{\mathcal{M}}$	$Var[D(\theta)]/2$
TUMOUR		T1			
β	β	β	β	6.51	6.77
$\beta(t)$	β	β	β	7.81	8.26
β	$\beta(t)$	β	β	7.46	7.41
β	β	$\beta(t)$	β	8.94	12.00
β	β	β	$\beta(t)$	7.39	7.92
$\beta(t)$	$\beta(t)$	β	β	8.65	9.39
β	$\beta(t)$	$\beta(t)$	β	10.24	11.74
β	$\beta(t)$	β	$\beta(t)$	8.95	9.49
$\beta(t)$	$\beta(t)$	$\beta(t)$	β	11.92	14.43
β	$\beta(t)$	$\beta(t)$	$\beta(t)$	11.64	13.28
$\beta(t)$	$\beta(t)$	$\beta(t)$	$\beta(t)$	12.77	15.14
—	$\beta(t)$	$\beta(t)$	β	10.08	11.54
β	—	$\beta(t)$	β	8.43	11.20
β	$\beta(t)$	—	β	6.53	6.97
β	$\beta(t)$	$\beta(t)$	—	9.31	11.22

slightly rises, when the fitted dynamic structure is more complex, i.e. when T1 is included with a time-varying effect.

Subsequently, the predictor of the selected multivariate model $\mathcal{M}^*$ is extended by including GRADING with a fixed effect. Again it was assumed that is does not provide additional information and hence only model complexity increases while the goodness of fit stays more or less unchanged. The values of the criteria for model $\mathcal{M}^*$ and the model including GRADING are given in the lower part of Table 6.4. The DIC and the measures involved in its determination support this assumption. The effective number of model parameters rises by 0.8 while the posterior mean of the deviance $\overline{D}$ measuring the fit of the model, declines indiscernibly. Consequently the DIC slightly increases, indicating that the goodness of fit is not sufficiently improved to justify the additional consideration of GRADING. In contrast, the approximate Bayes factor and the posterior Bayes factor both slightly decrease and obviously penalise the additional complexity less. Yet, the change in these criteria is by far too small to indicate 'strong evidence' for GRADING. In the last model of Table 6.4

the effect of GRADING is allowed to vary over time. Here all criteria (except of the model deviance $D(\bar{\theta})$) coincide by rejecting an additional dynamic effect.

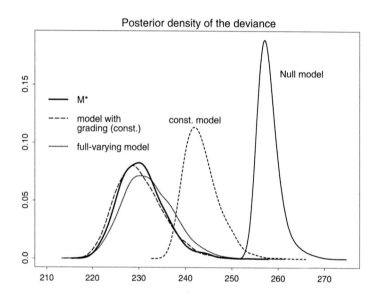

Figure 6.8: The posterior density of the deviance for different models.

As described in Section 4.4 an alternative method for the evaluation of the goodness of fit of a model is the comparison of the posterior distributions of the deviance $D(\theta)$. In Figure 6.8 density plots of those posterior distributions are displayed for the three static solutions, the optimal model $\mathcal{M}^*$ and the model including tumour grading with an time-constant effect. It shows that all models considering time-variation are clearly superior to the time-constant model and the null model. However Figure 6.8 also points out the difficulties in using these plots for a visual model selection process as used above. The models with a close fit are hard to distinguish.

6.4 Discussion

The focus of the presented analysis was not to develop a new prognostic system for breast cancer patients but on providing a comparative study of the approaches to dynamic survival modelling and model selection that have been presented in this thesis. From this point of view it is remarkable how close the results of these different approaches are.

The determination of the optimal effect structure in the Fractional Polynomial approach using the backfitting-type algorithm results in a semi-parametric predictor where the effects of the hormone receptor state and of the new protein T1 are allowed to change over time. Also, in the Bayesian approach the effects of these two factors are chosen to be dynamic, while for tumour size and MIB1 time-constant effects are selected when the Bayes factor, the posterior mean of the deviance $\overline{D}(\theta)$ and the Deviance Information criterion DIC are optimised. The posterior Bayes factor seems to put too much weight on the goodness of fit, and undercharges model dimension. Thus it shows equal values also for more complex models. The model deviance $D(\bar{\theta})$, as expected, generally declines with increasing model complexity and obviously does not take model dimension into account. Consequently it selects the most complex model. The DIC rather tends to more parsimonious models while the Bayes factor equally rates the model including an additional dynamic effect for tumour size as well as the model which only allows time-variation for the effects of hormone receptor state.

Figure 6.3 and Figure 6.7 show similar shapes for the resulting dynamic curves of the two approaches. The effect function of the hormone receptor state linearly increases over time, confirming findings that have been reported before (e.g. Berger et al., 1997). The effect of T1 follows a bent curve.

Note that the Fractional Polynomial approach selects global polynomial functions to smoothly describe the dynamic development of the effects. In contrast the second order random walk used in the Bayesian approach rather produces a sequence of $\beta(t)$'s, where smoothness it obtained by a local penalisation of deviations from the straight line. The courses of the dynamic effects for the two factors are flatter and have narrower error bands than in the Fractional Polynomial approach. By restricting the effects of tumour size and MIB1 to time-constancy in the Bayesian logit model, their error standard bands yield the same results as the Fractional Polynomial model.

It was shown in Section 2 that the logit model for interval censored survival times approximates the Cox model as the interval lengths approach zero. In this data set observation time is measured in months and are thus discrete. However, due to the rather low event rate tied event times occur in only three time-points. In addition, observation time is rather long, ranging from 10 to 127 months. Thus the data is approximately continuous and the logit model and the Cox model can be assumed to be approximately equivalent.

Some discrepancy between the approaches was observed when evaluating the global impact of the factors on variable selection. A particular focus was on tumour size and tumour grading. In the Fractional Polynomial approach tumour size clearly had no significant impact on disease free survival. In the Bayesian approach to evaluate the global impact of tumour size, the model omitting this factor $\mathcal{M}_{-TUMOUR}$ is compared with the optimal model $\mathcal{M}^*$ including tumour size with a constant effect. Following the proposed interpretation schemes given in Section 4.4 the Bayes factor yields strong evidence in favour for model $\mathcal{M}^*$ with $2\log(\text{BF}(\mathcal{M}_{-TUMOUR})) - 2\log(\text{BF}(\mathcal{M}^*)) = 3.33$. Also the posterior Bayes factor, the posterior mean of the deviance $\overline{D}$ and the Deviance Information criterion DIC all slightly increase when tumour size is omitted from the model. However, the difference to model $\mathcal{M}^*$ is less then 2, suggesting that there is no substantial loss in goodness of fit. Bearing in mind the approximative IC-representations of the criteria, these results surprise, as one would presume the Bayes factor to penalise complexity most. Moreover, the criteria are expected to decrease, as the posterior mean of the effect of TUMOUR is not very large ($\beta_{TUMOUR} = 0.598$) and the ± 2 posterior standard error bands even include zero, i.e. $[-0.353, 1.549]$.

When in addition to the four initial factors tumour grading is included into the model, its impact was again validated differently within the different approaches. The Fractional Polynomial approach found no significant improvement in fit when adding grading. When grading was included in the Bayesian logit model with a constant effect only the DIC showed a slight increase. All other criteria declined somewhat, although the improvement was not large enough to provide positive evidence in favour of the extended model. Since the extended model is at the same time more complex, inclusion of grading was rejected. A dynamic effect structure for grading was denied by all Bayesian model criteria.

These results give the impression that the Bayesian approach is more generous with selecting complex models. Yet, it should be noted that in the Fractional Polynomial approach the selected significance level of 0.05 must be achieved to accept a more complex model. If in Bayesian model selection 'strong evidence' is stipulated to select a model, higher complexity would only be accepted if the criteria decrease by more than 6. The Bayesian model selection algorithms based on the Bayes factor, the posterior Bayes factor, the Deviance and the DIC, which are illustrated in Table 6.5, would then have stopped already in the second step, deciding that beside the hormone receptor state no other factor requires dynamic effects.

Chapter 7

Summary

The exploration of time-variation in survival models can be summarised in two general questions: First, can the effect of a covariate be assumed to be time-constant, or does it distinctly change over time? Secondly, if the effect changes over time, of which form is the dynamic structure. In this thesis possible approaches to explore these questions have been investigated within two opposite concepts of statistics: Likelihood based statistics and Bayesian statistics. The fundamental differences of these two concepts led to different approaches when exploring dynamic effect structures: Model estimation in the likelihood framework was realised by maximising variants of the likelihood, while in Bayesian statistics inference is based on posterior distributions of the model parameters. In the likelihood framework model choice can be based on likelihood ratio testing, while in the Bayesian framework, Bayesian model criteria are employed to compare different, non-nested models.

Beside the theoretical framework, also the particular structure of the data at hand can lead to different modelling approaches, as survival time might either be continuous or discrete. Both situations have been considered in this thesis, employing two different models, the dynamic Cox model for continuous survival time and the dynamic logit model for discrete survival time.

Within the Cox model framework time-varying effects play a decisive role, since they lead to the violation of the underlying proportional hazards assumption. Most approaches found in literature, which cope with time-varying effects in the Cox model, either demand for prespecifications on the presumed underlying dynamic structure or require complex estimation procedures. In application these requirements are serious hurdles. In this thesis the Fractional polynomial approach for dynamic modelling

165

was introduced. It describes the temporal development of the effects by parametric functions but does not need any prespecifications. As a result model estimation can be based on standard estimation procedures readily available in most statistical software packages. To decide upon the inclusion of dynamic structures a likelihood ratio test was developed. This test based on Fractional Polynomials gives a flexible alternative for detecting complex departures of the proportional hazards assumption of the Cox model. In simulation studies the test proved to be consistent and showed omnibus power in different dynamic situations. In addition, an extension of this likelihood ratio test was defined to allow for variable selection. Bonferroni adjustment of the p-value permits a joint application of the two tests. This was used in an algorithm for fitting multivariate semi-dynamic Cox models, which automatically decides which of the covariate effects should be modelled in a dynamic fashion and which can be assumed to be constant. The application of the algorithm in a study of gastric cancer prognosis showed that the resulting effect functions constructed by Fractional Polynomials are sufficiently flexible to embrace underlying dynamic structures. Furthermore, the automatic model selection ensures that the resulting fit is as parsimoniously as the data allows.

The qualities of the proposed method of a) providing a flexible fit for dynamic effect structures, b) offering a reliable test, while c) being computationally straightforward, makes the Fractional Polynomial approach a promising tool for applied survival analysis.

Modelling of discrete survival data was performed within the Bayesian framework. The dynamic logit model was employed, where discrete survival data were regarded as a series of binary outcomes. Model inference in the Bayesian survival modelling context relies on posterior samples, which have been generated by Markov Chain Monte Carlo methods. To incorporate dynamic effects, hierarchical models of state space structures could be employed, where smooth curves for the dynamic effects are modelled by a second order random. The advantageous feature of this approach is that the random walk variance is estimated from the data simultaneously with the dynamic effects. Since it acts as a smoothing parameter, this can be regarded as a data-driven decision on the degree of smoothness of the fit.

A major focus in this part of the thesis was on Bayesian model choice, with particular emphasis placed on the assessment whether the fitted dynamic effects are actually required or not. In the Bayesian framework the analogue to significance testing is provided by the Bayes factor, which allows for a direct comparison of models with constant effects and dynamic models. The Bayes factor suffers, however, from serious limitations. Besides its sensitivity to prior specifications the major draw-back is its computation, which is of particular challenge in highly complex models, such as dynamic survival modes. A review of different proposals to determine the Bayes factor showed that within the context of non-parametric survival modelling most suggested routines are not practical. Only its approximation by the posterior harmonic mean proved to be feasible. In addition to the Bayes factor alternative 'posterior' criteria have been studied. An essential requirement was that they allow for comparison of non-nested models with arbitrary complexity. This was found to be valid for the posterior Bayes factor, the posterior expected deviance and the Deviance Information Criterion. Since all criteria can be derived directly from the MCMC-output their calculation is straightforward and requires negligible computational effort. The performance of these criteria has been investigated in a simulation study regarding different dynamic settings. The result of this comparative simulation study was of special interest, as all considered model criteria have been subject to controversial discussions within the Bayesian community. It revealed that the conclusions based on the different criteria do not fundamentally differ, at least not in the model selection problems regarded in this thesis.

The criteria were then employed to select a model appropriately describing the impact of different factors on infant mortality. Since the infant mortality data are clearly discrete, the logit approach was most suitable. The study gave an illustrative example for a complex modelling process and demonstrated, how generally applicable the considered criteria are. They allowed for the comparison of arbitrary complex, non-nested models, which included constant, dynamic and additive effects. Both, the simulation study and the application showed, however, that the estimate of the effective complexity proposed by Spiegelhalter et al. produces contradictory results when applied to non-parametric hierarchical models and seems to remain an issue for further investigations.

A discussion of the different concepts of the criteria and a review of criticism found in literature lead to a more detailed insight into the theoretical backgrounds and allowed to point out disagreements of the concepts. In particular the Deviance Information criterion was illuminated and brought in context to the posterior distribution of the deviance, which yielded a justification for its construction. Furthermore, despite of their dissimilar theoretical backgrounds and their different justifications, the considered Bayesian criteria were shown to be related to each other on a common ground. This in turn offered the possibility of a better understanding of their behaviour in applications. The Bayes factor approximation, the posterior Bayes factor and the posterior expected deviance could be represented as different posterior

means of the likelihood, which gives a ranking of these criteria. In addition, large
sample properties allowed for all criteria to define an approximation, which had the
structure of information criteria consisting of the maximum likelihood deviance as
a measure of the goodness of model fit penalised by some multiple of the model dimension. This in turn provided insight into the different weights that each criterion
puts on penalisation of model complexity. It was seen that these weights are in
accordance with the order revealed by the mean-representation.

Finally, in a joint application on breast cancer data the two approaches have been
compared. Time was given in months, i.e. on a discrete scale, and thus a logit
model was employed. Due to the long observation time and a small number of tied
events the data also allowed using a Cox model. This comparative application was of
special interest since the two approaches fundamentally differ in various theoretical
aspects. The key issue is of course the underlying controverse statistical concepts of
the likelihood based approach and the Bayesian approach. This especially effects the
model selection process, that is the comparison of dynamic versus constant effects.
Moreover, the two approaches also differ by technical aspects of inference. While
Fractional Polynomials give global solutions for the dynamic effect functions, the
random walk prior used in the logit model locally smoothes roughness. The joint
application reveals that in the considered data situation, despite of their different
background, the outcome of the Fractional Polynomial approach and the Bayesian
approach are very similar. This indicates that both methods yield reliable results.
Of course the conclusions of this comparative study do not suggest that the two
approaches are generally interchangeable.

So far we have not mentioned the issue of coping with medium sized or small data
sets, which often is a relevant aspect in practical data analysis. Theoretically both
approaches promise to handle such data situations satisfactory. The Fractional Polynomial approach is rather parsimonious as it describes the dynamic development of
the effects by parametric functions that require only few parameters. Bayesian approaches are generally known to provide adequate results when fitting smaller data
sets as they incorporate prior informations. For the breast cancer analysis only a
small number of independent observations was available and it was shown that both
approaches still yield convincing results.

Outlook

All approaches have their limitations. Although Fractional Polynomials offer a reliable tool to fit dynamic effect structures within the Cox framework for some situations their global functions will not be flexible enough to ensure a good fit. This applies in particular for sudden changes or unexpected patterns in the dynamic effect structure, which might be overseen by global functions. In this situation local solutions might yield better results. Thus tests that allow for the comparison of parametric fit versus non-parametric fit within the Cox framework are needed. Royston (2000b) proposed such a test based on smoothing splines. Yet, the results of the gastric cancer study gave the impression, that smoothing splines are subject to artefacts which may in particular occur in small survival data sets. Possible alternatives for further research might be given by local likelihood fitting, as e.g. introduced by Kauermann & Berger (2002).

For model selection within the Bayesian framework, besides the model criteria discussed in this thesis, alternative proposals have been suggested, which rely on replicate data minimising posterior loss functions (see Gelfand & Gosh, 1998). These approaches might offer an attractive alternative that circumvents the problem of using the data twice. Since replicates can be generated simultaneously to posterior sampling, it does not require an overly computational effort. However, generating replicate survival data in the presence of censored observations is not straightforward.

Generally, due to the peculiarity of survival data, interpretation of dynamic effect structures need to be done with caution, eliminating any possibility of artefacts. Model selection should therefore in any case never be left to automatic algorithms alone, but should always additionally incorporate expert's knowledge.

Appendix A

Bayesian Model Assumptions

In the following the independence assumptions are given, that are used in Bayesian dynamic logit model (Fahrmeir & Lang, 2001). They allow to simplify the joint distribution of the parameters and are of particular benefit when constructing a MCMC algorithm for posterior sampling.

Let α_t^* denote the history of the state vector up to time t, so that $\alpha_t^* = (\alpha_1, \ldots, \alpha_t)'$. Accordingly let $y^*(t)$ denote the survival experience up to time t, i.e. $y^*(t) = (y(1), \ldots, y(t))'$. For the definition of a dynamic logit model with state-space structure, where the transition model of the states has the form of a random walk of order κ, the following independence assumptions hold:

A1 Conditional on the covariates x_n, the preceding survival history $y^*(t-1)$ and actual state α_t, $y_n(t)$ is independent of the former states α_{t-1}^* and the transition-variance Σ^2, i.e.

$$P\big(y_n(t)|y_n^*(t-1), x_n, \alpha_t^*, \Sigma^2\big) = P\big(y_n(t)|y_n^*(t-1), x_n, \alpha_t\big).$$

A2 Within a risk set R_t the individual event indicators $y_n(t)$ are conditional independent, i.e. for $y(t) = y_1(t), \ldots, y_N(t)$ and $x = (x_1, \ldots, x_N)$

$$P\big(y(t)|y^*(t-1), x, \alpha_t\big) = \prod_{n \in R_t} P(y_n(t)|y_n^*(t-1), x_n, \alpha_t).$$

171

A3 The covariates and the censoring process are conditional independent of past states α_{t-1}^*, as generally assumed in survival analysis.

A4 The sequence of states $\alpha_1, \ldots, \alpha_k$ is a Markov chain of order κ, i.e.

$$P(\alpha_t | \alpha_{t-1}^*, \Sigma^2) = \begin{cases} P(\alpha_t | \alpha_{t-1}, \ldots, \alpha_{t-\kappa}, \Sigma^2) & t > \kappa \\ P(\alpha_t) & t \leq \kappa \end{cases}.$$

Furthermore, the initial values $\alpha_1, \ldots, \alpha_\kappa$ and Σ^2 are mutually independent. This corresponds to the general state space assumption.

Appendix B

Implementation of the Algorithms

As part of this thesis the proposed methods have been implemented for the application to exemplifying datasets and explorative simulation studies. In addition, various alternative proposals have been implemented to perform comparative studies.

The multivariate algorithm for fitting semi-dynamic Cox models based on the Fractional Polynomial approach was implemented in S^+, where an add-on library was created, which can be integrated in any S^+-session. It is available on request. The procedures of the library allow to fit univariate as well as multivariate dynamic Cox models, where single covariates can be forced to be fitted with a time-constant effect by the user, when semi-parametric models are required. As the Fractional Polynomial approach can be based on standard estimation techniques of the Cox model, implementation was based on the coxph-function of S^+, where I used the representation of the time-dependent covariate model as a stratified Cox model. This allowed a fast computation of the Fractional Polynomial models. The use of the procedures is straightforward and allows the adaptation of the power-set to the actual modelling situation. In addition, the significance level for the PH-test can be changed as well as other parameters of the algorithm. An additional procedure graphically illustrates the results.

The FP method was compared to various other proposals for fitting and testing dynamic effect structures within the Cox framework. This comparison was performed in S^+, too. I partly used ready available procedures, e.g. for the residual score test of Grambsch & Therneau (cox.zph) or for Hastie & Tibshirani's smoothing spline fit (S^+-library by the authors, which can be found as shared software at http://lib.stat.cmu.edu/S/), other methods I have implemented, as e.g. Hess's regression splines and the test based on a piece-wise constant fit. The test on violation of the PH-assumption proposed by Cox can be determined using the FP-library by appropriately specifying the power-set.

To fit Bayesian dynamic logit models for survival data a Markov Chain Monte Carlo sampling algorithm was implemented in C^{++}. This allows for a fast generation of relatively huge posterior samples, where the algorithm efficiently takes the special structure of survival data into account. The algorithm was used for the simulation study and the analysis of the breast cancer data. It includes the calculation of all important posterior quantities and returns the values for all proposed model criteria. Additionally the program BayesX of Lang & Brezger (2001b) was used, a shared software package that was created within the Sonderforschungsbereich SFB386. BayesX allows for MCMC-estimation of a variety of generalised linear models with different effect structures, such as additive effects, varying and dynamic effects, random and spatial effects ect. Additional S^+-functions allow to plot the MCMC-output for further analysis. BayesX was used to analyse the infant mortality data, where in one of the models an additive effect was included. However, to fit survival data within BayesX, the data set has to be restructured appropriately. Thereto a procedure was written in S^+, which transforms common survival data into a binary time series.

All programs and procedures are available from the author upon request: (`ursula.berger@imse.med.tu-muenchen.de`).

Bibliography

[1] Aitkin, M. (1991). Posterior Bayes factor, (with discussion). *Journal of the Royal Statistical Society*, B, 53, pp. 111-142.

[2] Aitkin, M. (1997). The calibration of P-values, posterior Bayes factors and the AIC from the posterior distribution of the likelihood, (with discussion). *Statistics and Computing*, 7, pp. 253-272.

[3] Akaike, H. (1973). Information theory and an extension of the maximum likelihood principle. *Proceedings of the 2nd Int. Symposium on Information Theory*, Budapest, pp. 267-281.

[4] Akaike, H. (1991). Discussion of the paper by Aitkin. *Journal of the Royal Statistical Society*, B, 53, p. 135.

[5] Andersen, P.K., Borgan, O., Gill, R., Keiding, N. (1993). *Statistical models based on counting processes*. Springer-Verlag, New York.

[6] Berger, J. (1985). *Statistical decision theory and Bayesian analysis*. 2nd edition, Springer-Verlag, New York.

[7] Berger, J., Pericchi, L. (1996). The intrinsic Bayes factor for model selection and prediction. *Journal of the American Statistical Association*, 91, pp. 109-122.

[8] Berger, J., Pericchi, L. (2001). Objective Bayesian methods for model selection: Introduction and comparison, (with discussion). In: *Model Selection* (Ed.: P. Lahiri), Institute of Mathematical Statistics Lecture Notes – Monograph Series, 38.

[9] Berger, U., Ulm, K., Harbeck, N., Schmitt, M. (1997). Identifying time-varying effects of prognostic factors in breast cancer patients. *Breast Cancer Research and Treatment* (Ed.: M. Lippman), Kluwer Academic Publishers, 41.

[10] Berger, U., Gerein, P., Ulm, K., Schäfer, J. (2000). On the use of Fractional Polynomials in dynamic Cox models. SFB 386: Discussion Paper 207, Ludwig-Maximilians-Universität München.

Published as

Berger, U., Ulm, K., Schäfer, J. (2003). Dynamic Cox modelling based on fractional polynomials: time-variations in gastric cancer prognosis. *Statistics in Medicine*, 22, pp. 1163-1180.

[11] Bernardo, J., Smith, A. (1993). *Bayesian theory*. John Wiley, Chichester.

[12] Biller, C. (2000). Adaptive Bayesian regression splines in semiparametric generalized linear models. *Journal of Computational and Graphical Statistics*, 9, pp. 122-140.

[13] Breslow, N.E. (1974). Covariance analysis of censored survival data. *Biometrics*, 30, pp. 89-99.

[14] Carlin, B., Chib, S. (1995). Bayesian Model Choice via Markov Chain Monte Carlo Methods. *Journal of the Royal Statistical Society*, B, 57, pp. 473-484.

[15] Chib, S. (1995). Marginal likelihood from the Gibbs output. *Journal of the American Statistical Association*, 90, pp. 1313-1321.

[16] Chib, S., Jeliazkov, I. (2001). Marginal likelihood from the Metropolis-Hastings output. *Journal of the American Statistical Association*, 96, pp. 270-281.

[17] Cox, D.R. (1972). Regression models and life-tables, (with discussion). *Journal of the Royal Statistical Society*, B, 34, pp. 187-220.

[18] Cox, D.R. (1975). Partial likelihood. *Biometrika*, 62, pp. 269-276.

[19] Cox, D.R., Oakes, D. (1984). *Analysis of survival data*. Chapman and Hall, London.

[20] DeGroot, M. (1982). Comment. *Journal of the American Statistical Association*, 77, pp. 336-338.

[21] Dempster, A. (1997a). The direct use of likelihood for significance testing, (with discussion). *Statistics and Computing*, 7, pp. 247-252. (Reprint of 1974).

[22] Dempster, A. (1997b). Comment on 'The calibration of *P*-values, posterior Bayes factors and the AIC from the posterior distribution of the likelihood'. *Statistics and Computing*, 7, pp. 265-269.

[23] Draper, D. (1995). Assessment and propagation of model uncertainty, (with discussion). *Journal of the Royal Statistical Society*, B, 57, pp. 45-97.

[24] Efron, B. (1977). The efficiency of Cox's likelihood function for censored data. *Journal of the American Statistical Association*, 72, pp. 557-565.

[25] Efron, B. (1988). Logistic regression, survival analysis, and the Kaplan-Meier curve. *Journal of the American Statistical Association*, 83, pp. 414-425.

[26] Fahrmeir, L. (1994). Dynamic modelling and penalized likelihood estimation for discrete time survival data. *Biometrika*, 81, 2, pp. 317-330.

[27] Fahrmeir, L., Knorr-Held, L. (1997). Dynamic discrete time duration models: Estimation via Markov Chain Monte Carlo. *Sociological Methodology*, pp. 417-452.

[28] Fahrmeir, L., Lang, S. (2001). Bayesian inference for generalized additive mixed models based on Markov random field priors. *Applied Statistics*, 50, 2, pp. 201-220.

[29] Fahrmeir, L., Tutz, G. (2001). *Multivariate statistical modelling based on generalized linear models* (2nd edn.). Springer-Verlag, New York.

[30] Fleming, T.R., Harrington, D.P. (1991). *Counting processes and survival analysis.* Wiley Series, New York.

[31] Gelfand, A., Deya, D., Cahng, H. (1992). Model determination using predictive distributions with implementation via sampling-based methods, (with discussion). In: *Baysian Statistics 4* (Ed.: J. Bernardo, J. Berger, A. Dawid, A. Smith). Oxford University Press, Oxford, pp. 147-167.

[32] Gelfand, A., Ghosh, S. (1998). Model choice: A minimum posterior predictive loss approach. *Biometrika*, 85, 1, pp. 1-11.

[33] Gelfand, A., Ghosh, S. (2001). Generalized linear models: A Bayesian view. In: *Generalized linear models: A Bayesian perspective* (Ed.: D. Dey, S. Ghosh, B. Mallick). Marcel Dekker, New York.

[34] Gelman, A. (1996). Inference and monitoring convergence. *Markov Chain Monte Carlo in practice* (Ed.: W. Gilks, S. Richardson, D. Spiegelhalter). Chapman and Hall, London, pp. 131-144.

[35] Gilks, W., Richardson, S., Spiegelhalter, D. (1996). *Markov Chain Monte Carlo in practice.* Chapman and Hall, London.

[36] Gore, S.M., Pocock, S.J., Gillian, R.K. (1984). Regression models and non-proportional hazards in the analysis of breast cancer survival. *Applied Statistics*, 33, 2, pp. 176-195.

[37] Grambsch, P., Therneau, T.M. (1994). Proportional hazards tests and diagnostics based on weighted residuals. *Biometrika*, 81, 3, pp. 515-526.

[38] Gray, R.J. (1992). Flexible methods for analyzing survival data using splines, with application to breast cancer prognosis. *Journal of the American Statistical Association*, 87, pp. 942-951.

[39] Gray, R.J. (1994). Spline-based test in survival analysis. *Biometrics*, 50, pp. 640-652.

[40] Green, P. (1995). Reversible jump Markov Chain Monte Carlo computation and Bayesian model determination. *Biometrika*, 57, pp. 97-109.

[41] Han, C., Carlin B.P. (2000). MCMC methods for computing Bayes factors: A comparative review. Research Report 2000-001, Division of Biostatistics, University of Minnesota.

[42] Harbeck, N., Dettmar, P., Thomssen, C., Berger, U., Ulm, K., Kates, R., Höfler, H., Jänicke, F., Graeff, H., Schmitt, M. (1999). Risk-group discrimination in node-negative breast cancer using invasion and proliferation markers: 6-year median follow-up. *British Journal of Cancer*, 80, 3-4, pp. 419-426.

[43] Harbeck, N., Alt, U., Berger, U., Kates, R., Kruger, A., Thomssen, C., Jänicke, F., Graeff, H., Schmitt, M. (2000). Long-term follow-up confirms prognostic impact of PAI-1 and cathepsin D and L in primary breast cancer. *International Journal of Biological Markers*, 15(1), pp. 79-83.

[44] Harrell, F. (2001). *Regression Modeling Strategies: With Applications to Linear Models, Logistic Regression, and Survival Analysis*, Springer-Verlag, New York.

[45] Harrell, F.E., Lee, K.L. (1986). Verifying assumptions of the Cox proportional hazards model. it SUGI II: Proceedings of the Eleventh Annual SAS Users Group International Conference, pp. 823-828.

[46] Hastie, T.J., Tibshirani, R.J. (1990). *Generalized additive models*. Chapman and Hall, London.

[47] Hastie, T.J., Tibshirani, R.J. (1993). Varying-coefficient models, (with discussion). *Journal of the Royal Statistical Society*, B, 55, 4, pp. 757-796.

[48] Hess, K.R. (1994). Assessing time-by-covariate interactions in proportional hazards regression models using cubic spline functions. *Statistics in Medicine*, 13, pp. 1045-1062.

[49] Hess, K.R. (1995). Graphical methods for assessing violations of the proportional hazards assumption in Cox regression. *Statistics in Medicine*, 14, pp. 1707-1723.

[50] Jeffreys, H. (1961). *Theory of probability* (3rd edn.). Oxford University Press, London. (reprint of 1939)

[51] Kalbfleisch, J.D., Prentice, R.L. (1980). *The statistical analysis of failure time data*. Wiley, New York.

[52] Kass, R., Raftery, A. (1995). Bayes factors. *Journal of the American Statistical Association*, 90, pp. 773-795.

[53] Kauermann, G., Berger, U. (2002). A Smooth Test in Proportional Hazard Survival Models Using Local Partial Likelihood Fitting. *Lifetime Data Analysis*, to appear.

[54] Knorr-Held, L. (1997). *Hierarchical modelling of discrete longitudinal data - applications of Markov Chain Monte Carlo.* Herbert Utz Verlag, München.

[55] Knorr-Held, L. (1999). Conditional prior proposals in dynamic models. *Scandinavian Journal of Statistics*, 26, pp. 129-144.

[56] Kupper, L.L., Stewart, J.R., Williams, K.A. (1976). A note on controlling significance levels in stepwise regression. *American Journal of Epidemiology*, 103, 1, pp.13-15.

[57] Lang, S., Brezger, A. (2001a). Bayesian P-splines. Discussion paper 236, SFB 386, Department of Statistics, Ludwig-Maximilians-Universität München.

[58] Lang, S., Brezger, A. (2001b). BayesX, Software for Bayesian inference based on Markov Chain Monte Carlo simulation techniques. Department of Statistics, Ludwig-Maximilians-Universität München.

[59] Lindely, D. (1991). Discussion of the paper by Aitkin. *Journal of the Royal Statistical Society*, B, 53, p. 130.

[60] Lindely, D. (1957). A statistical paradox. *Biometrica*, 44, pp. 187-192.

[61] McCullagh, P., Nelder, J. A. (1983). *Generalized linear models.* Chapman and Hall, London.

[62] Moreau, T., O'Quigley, J., Mesbah, M. (1985). A global goodness-of-fit statistic for the proportional hazard model. *Applied Statistics*, 34, 3, pp. 221-218.

[63] Moreau, T., O'Quigley, J., Lellouch, J. (1986). On D. Schoenfeld's approach for testing the proportional hazards assumption. *Biometrika*, 73, 2, pp. 513-515.

[64] Nekarda, H., Schmitt, M., Ulm, K., Wenninger, A., Vogelsang, H., Becker, K., Roder, J.D., Fink, U., Siewert, J.R. (1994). Prognostic impact of urokinase-type Plasminogen Activator uPA and its inhibitor PAI-1 in gastric cancer with complete resection (R0-Category, UICC). *Cancer Research*, 54, pp. 2900-2907.

[65] Newton, M., Raftery, A. (1994). Approximate Bayesian inference with the weighted likelihood bootstrap, (with discussion). *Journal of the Royal Statistical Society*, B, 56, pp. 3-48.

[66] Ng'Andu, N.H. (1997). An empirical comparison of statistical tests for assessing the proportional hazards assumption of Cox's model. *Statistics in Medicine*, 16, pp. 611-626.

[67] O'Hagan, A. (1991). Discussion of the paper by Aitkin. *Journal of the Royal Statistical Society*, B, 53, p. 136.

[68] O'Hagan, A. (1995). Fractional Bayes Factors for model comparison. *Journal of the Royal Statistical Society*, B, 57, pp. 99-118.

[69] Parmar, M. and Machin, D. (1996). *Survival Analysis*, John Wiley, New York.

[70] Prechtel, D., Harbeck, N., Berger, U., Höfler, H., Werenskiold, A.K. (2001). Clinical relevance of T1-S, an oncogene-inducible, secreted glycoprotein of the immunoglobulin superfamily, in node-negative breast. *Laboratory Investigation*, 81(2), pp. 159-65.

[71] Ripley, B. (1987). *Stochastic simulation*. John Wiley, New York.

[72] Royston, P. (2000a). A useful monotonic non-linear model with applications in medicine and epidemiology. *Statistics in Medicine*, 19, pp. 2053-2066.

[73] Royston, P. (2000b). A strategy for modelling the effect of continuous covariates in medicine and epidemiology. *Statistics in Medicine*, 19, pp. 1831-1847.

[74] Royston, P. and Altman, D.G. (1994). Regression using Fractional Polynomials of continuous covariates: Parsimonious parametric modelling. *Applied Statistics*, 43, pp. 429-467.

[75] Sargent, D.J. (1997). A flexible approach to time-varying coefficients in the Cox regression setting. *Lifetime Data Analysis*, 3, pp. 13-25.

[76] Sauerbrei, W., Royston, P. (1999). Building multivariable prognostic and diagnostic models: Transformation of the predictors by using Fractional Polynomials. *Journal of the Royal Statistical Society*, B, 162, pp. 77-94.

[77] Schoenfeld, D. (1980). Chi-squared goodness-of-fit tests for the proportional hazards regression model. *Biometrika*, 67, 1, pp. 145-153.

[78] Schoenfeld, D. (1982). Partial residuals for the proportional hazards regression model. *Biometrika*, 69, 1, pp. 239-241.

[79] Schwarz, G. (1978). Estimating the dimension of a model. *Annals of Statistics*, 6, pp. 461-464.

[80] Shafer, G. (1982). Lindley's paradox, (with discussion). *Journal of the American Statistical Association*, 77, pp. 325-351.

[81] Siewert, R., Bottcher, K., Stein, H.J., Roder, J.D. (1998). Relevant prognostic factors in gastric cancer: Ten-year results of the German Gastric Cancer Study. *Annals of Surgery*, 228, 4, pp. 449-461.

[82] Smith, A.F.M., Spiegelhalter, D.J. (1980). Bayes Factors and Choice Criteria for Linear Models. *Journal of the Royal Statistical Society*, B, 42, pp. 213-220.

[83] Spiegelhalter, D.J., Smith, A.F.M. (1982). Bayes factors for linear and loglinear models with vague prior information. *Journal of the Royal Statistical Society*, B, 44, pp. 377-387.

[84] Spiegelhalter, D.J., Best, N.G., Carlin, B.P., van der Linde, A. (2001). Bayesian measures of model complexity and fit. Research Report 2001-013, Division of Biostatistics, University of Minnesota.
Published as:
Spiegelhalter, D.J., Best, N.G., Carlin, B.P., van der Linde, A. (2002). Bayesian measures of model complexity and fit, (with discussion). *Journal of the Royal Statistical Society*, B, 64, pp. 583-639.

[85] Stablein, D.N., Carter, W.H.Jr., Novak, J.W. (1981). Analysis of survival data with nonproportional hazard functions. *Controlled Clinical Trials*, 2, pp. 149-159.

[86] Statistisches Bundesamt (1994). *Statistisches Jahrbuch für die Bundesrepublik Deutschland*. Statistisches Bundesamt, Wiesbaden.

[87] Stone, M. (1997). Discussion of papers by Dempster and Aitkin. *Statistics and Computing*, 7, pp. 263-264.

[88] Thompson, W.A. (1977). On the treatment of grouped observations in life studies. *Biometrics*, 33, pp. 463-470.

[89] Verweij, P., van Howelingen, H. (1995). Time-dependent effects of fixed covariates in Cox regression. *Biometrics*, 51, pp. 1550-1556.

[90] WHO (1995). Physical Status: The use and interpretation of anthropometry. WHO Technical Report Series, 854, Geneva.

[91] World Bank (1995). Zambia poverty assessment. The World Bank, Washington, DC.

[92] Zucker, D. and Karr, A. (1990). Non-parametric survival analysis with time-dependent covariate effects: A penalized likelihood approach. *Annals of Statistics*, 18, pp. 329-352.

Lebenslauf

Name:		Ursula Berger
Geboren:	3. März 1970	in München

Schulbildung:	1976 – 1980	Berner Grundschule, München
	1980 – 1982	Thomas-Mann Gymnasium, München
	1982 – 1988	Europäische Schule München

Studium:	ab WS 1988/1989	Studium der Statistik an der Ludwig-Maximilians-Universität München
	Dezember 1994	Diplom in Statistik

Promotion:	Februar 2002	zum Dr. oec. publ. an der Ludwig-Maximilians-Universität München

Beruf:	Mai 1995 – Dezember 1996	Wissenschaftliche Mitarbeiterin an der Fakultät für Wirtschafts- und Sozialwissenschaften der Universität Potsdam
	seit Januar 1997	Wissenschaftliche Mitarbeiterin am Institut für Medizinische Statistik und Epidemiologie, Klinikum Rechts der Isar der Technischen Universität München